Is this Cell a Human Being?

Antoine Suarez • Joachim Huarte
Editors

Is this Cell a Human Being?

Exploring the Status of Embryos, Stem Cells and Human-Animal Hybrids

Editors
Dr. Antoine Suarez
The Institute for Interdisciplinary Studies
Berninastr. 85
8057 Zurich, Switzerland
suarez@leman.ch

Dr. Joachim Huarte
University of Geneva Medical School
Department of Genetic Medicine
and Development
1, rue Michel-Servet
1211 Geneva 4, Switzerland
joachim.huarte@gmail.com

ISBN 978-3-642-20771-6 e-ISBN 978-3-642-20772-3
DOI 10.1007/978-3-642-20772-3
Springer Heidelberg Dordrecht London New York

Library of Congress Control Number: 2011932848

Printed on acid-free paper

Springer is part of Springer Science+Business Media (www.springer.com)

Preface

Is this cell entity a human being? The answer to this question is crucial for making bioethical decisions either for or against research and experimentation.

In 1993, I proposed the developmental potential for spontaneous movements as a basis for determining whether a cell entity is a human being. Subsequently, Joachim Huarte and I implemented the spontaneous movement criterion for moral status in different contexts: patients in a persistent vegetative state; children with anencephaly; brain dead organisms; and cell entities arising from parthenogenesis. Over the same period, research advanced, bearing challenging new achievements: The production of so-called ng-parthenotes (which reach fetal motility and even birth) by fertilization-like procedures; the proposal of altered nuclear transfer; the reprogramming of somatic adult cells to induced pluripotent stem cells; and the possibility of changing pluripotent stem cells into an entire living animal using tetraploid complementation. These achievements increasingly suggested that the question of deciding whether a cell entity is a disabled human embryo or a non-embryo required stronger collaboration among experts in the field.

Thus, I conceived the idea of gathering interested scholars to debate the issue. Carlos Cavallé, the president of the Social Trends Institute (STI), immediately grasped the relevance of the project and generously agreed to sponsor and organize an Experts Meeting in Barcelona in January 2009. The presentations and debates from that encounter have been updated and complemented by additional articles to provide the content for this book.

Is this cell entity a human being? The question cries out for an answer, which this book attempts to provide. Experts in the field discuss the production of embryonic-like pluripotent stem cells by altered nuclear transfer, parthenogenesis, and reprogramming of adult somatic cells. They thoroughly analyze the biological and moral status of different cell entities, such as human stem cells, embryos, and human–animal hybrid embryos. The contributors do not always agree about the moral status of the cell entities under study. Each accepts responsibility only for the conclusions he or she draws.

Discussing these points of disagreement is crucial to finding out the correct answer. Thus, I am convinced that the ensemble of arguments presented in the book represent a decisive step toward establishing final criteria for determining what constitutes a human being, although a considerable amount of work – including new experiments – has still to be done.

I am especially indebted to Neville Cobbe and Joachim Huarte for their collaboration during the whole genesis of the project and in particular for assisting with the editing work. I thank STI president Carlos Cavallé and STI staff members Christa Byker and Craig Iffland for sponsoring and organizing the Barcelona Experts Meeting, as well as all the participants at the meeting and contributors to the volume. I thank also the Swiss Society of Bioethics for supporting the revising work. Finally, I acknowledge the enjoyable collaboration with Andrea Pillmann and Andrea Schlitzberger from Springer during the production of the book.

Zürich, Switzerland Antoine Suarez
July 2011

STI

Social Trends Institute

The Social Trends Institute (STI) is an international research center dedicated to the analysis of globally significant social trends. STI focuses its research on four subject areas: Family; Bioethics; Culture and Lifestyles; and Corporate Governance.

STI organizes Experts Meetings on specific topics within one of the research branches. These meetings are intended to foster open, intellectual dialogue between scholars from all over the world and from different academic backgrounds and disciplines. The scholars meet to present and discuss their original research papers in an academic forum. These papers, reviewed, and edited in light of the conference discussion form the basis of publications such as this one.

STI's sole aim is to promote research and scholarship of the highest academic standard. In doing so, it hopes to make a scholarly contribution toward understanding the varying and complex social trends that are intertwined with the modern world. STI is therefore committed to the ideas that make such scholarship possible: intellectual freedom, openness to a diversity of viewpoints, and a shared commitment to serve humanity.
www.socialtrendsinstitute.org

Contents

Contributors

Nicanor Pier Giorgio Austriaco, O.P. Department of Biology, Providence College, Providence, RI, USA, naustria@providence.edu

Neville Cobbe University of Liverpool, The Biosciences Building, Crown Street, Liverpool L69 7ZB, UK, Neville.Cobbe@liv.ac.uk

Maureen L. Condic Department of Neurobiology and Pediatrics, University of Utah School of Medicine Eccles Institute of Human Genetics, Bldg. 533, Room 2280, 15 North 2030 East, Salt Lake City, UT 84112–5330, USA, mlcondic@neuro.utah.edu

Boris Greber Department of Cell and Developmental Biology, Max Planck Institute for Molecular Biomedicine, Röntgenstraße 20, 48149 Münster, Germany

Joachim Huarte Department of Genetic Medicine and Development, University of Geneva Medical School, 1, rue Michel-Servet, 1211 Geneva 4, Switzerland (At the time of manuscript summission); Social Trends Institute/Bioethics, Barcelona, Spain; Swiss Society of Bioethics, Zurich, Switzerland, joachim.huarte@gmail.com

William B. Hurlbut Department of Neurology and Neurological Sciences, Stanford University Medical Center, Stanford, CA, USA, ethics@stanford.edu

Patrick Lee Franciscan University of Steubenville, Steubenville, OH, USA, plee@franciscan.edu, plee512@gmail.com

Pablo Requena Meana Pontifical University of the Holy Cross, Rome, Italy, requena@pusc.it

Hans Schöler Department of Cell and Developmental Biology, Max Planck Institute for Molecular Biomedicine, Röntgenstraße 20, 48149 Münster, Germany, office@mpi-muenster.mpg.de

Manfred Spieker University of Osnabrück, Osnabrück, Germany, mspieker@uni-osnabrueck.de

Antoine Suarez The Institute for Interdisciplinary Studies, Berninastr. 85, 8057 Zurich, Switzerland; Social Trends Institute/Bioethics, Barcelona, Spain; Swiss Society of Bioethics, Zurich, Switzerland, suarez@leman.ch

Chapter 1
Introduction

Joachim Huarte and Antoine Suarez

Abstract The search for principles allowing us to decide whether a particular cell entity is a human being or not, is the main motivation of this book. Even if the different contributors don't always reach a consensus in their proposals, the arguments they provide represent a significant step towards a consistent answer to the query.

Keywords Altered nuclear transfer • Human embryo • Moral status • Parthenogenesis • Reprogramming of somatic cells

The central question to which this book offers a series of answers is whether or not particular cell entities of human origin ought to be considered a human being. Answering this question is important insofar as one believes that there is some kind of intrinsic relationship between *human* species membership and the rational foundation of rights. On this view, the identification of a cell entity as an animal of the species *Homo sapiens* is relevant for making moral decisions vis-à-vis research and experimentation.

J. Huarte
Department of Genetic Medicine and Development, University of Geneva Medical School, 1, rue Michel-Servet, 1211 Geneva 4, Switzerland (At the time of manuscript summission)

Social Trends Institute/Bioethics, Barcelona, Spain

Swiss Society of Bioethics, Zurich, Switzerland
e-mail: joachim.huarte@gmail.com

A. Suarez (✉)
The Institute for Interdisciplinary Studies, Berninastr. 85, 8057 Zurich, Switzerland

Social Trends Institute/Bioethics, Barcelona, Spain

Swiss Society of Bioethics, Zurich, Switzerland
e-mail: suarez@leman.ch

A. Suarez and J. Huarte (eds.), *Is this Cell a Human Being?*,
DOI 10.1007/978-3-642-20772-3_1, © Springer-Verlag Berlin Heidelberg 2011

This question of identification became especially relevant in 2004 when many prominent research scientists proposed methods for deriving embryonic-like pluripotent stem cells without destroying human embryos. Such a solution had certain scientific and political advantages. On the one hand, such techniques only required human ova, not human embryos. This meant that researchers would not be restricted to solely using human embryos leftover from fertility clinics. On the other hand, such techniques, in theory, would not require the destruction of human embryos and would therefore bypass the political impasse over the subject of embryo experimentation.

The two main proposed methods were Altered Nuclear Transfer (ANT) and parthenotes. In ANT, one genetically alters the nucleus of a somatic cell (a skin cell, for example) before transferring it into an enucleated oocyte. In fact one inactivates a gene crucial for trophectoderm (TE) development. The inactivation eliminates the potential to form the fetal–maternal interface, but spares the inner cell mass (ICM) lineage. The resulting entity does not develop a healthy TE and is therefore incapable of uterine implantation. Researchers then would use the ICM of this entity to produce pluripotent stem cells sharing the self-renewal and differentiation potential of embryonic stem cells (ESCs). A parthenogenetic blastocyst, on the other hand, results from the development of an egg that only possesses a maternal genome, whose resultant ICM can also be used to derive pluripotent stem cells. Given these proposals, it was important from an ethical point of view, but was also interesting from a scientific point of view, to ascertain whether these alternative techniques produce cell entities that would be more properly classified as disabled human beings or if, indeed, they are not human beings at all. The ensuing debate raised interesting ethical and scientific questions, but also proved to be a profound cultural moment in which scientists and moral philosophers collaborated extensively with one another in order to get to the heart of the matter.

Four years later, however, the interest prompted by this discussion wavered due to the arrival of induced pluripotent stem cells (iPSCs) obtained through the reprogramming of adult somatic cells. Also this method made it possible to produce embryonic-like pluripotent stem cells without the necessity of destroying embryos and the prevailing consensus was that iPSCs were not human beings. However, recent research casts doubts about whether the reprogramming of adult cells could become a new sort of cloning. Indeed, it is already possible to create a mouse fully derived from iPSCs. What this suggests is the possibility that iPSCs could be considered equivalent to the ICM of a developing human embryo. Suppose that by improving the reprogramming methods one obtains "better" iPSCs. Suppose that one injects these "better" iPSCs into a TE vesicle from which the ICM cells have been removed. Suppose that this reconstituted blastocyst in which the ICM has been replaced by the iPSCs is capable of developing to birth. If this is the case, then we need to ask whether or not the destruction of a cluster of such "better" iPSCs could be equivalent to the destruction of a human embryo.

In order to answer this question, one is required to think about the necessary epigenetic components and functionality required for scientists to consider a certain cell entity to be a human being. Attempting to identify those components requires scientific knowledge and philosophical precision since one will be required to identify which components are directly or *essentially* related to the identity of a human person and which components are indirectly or *accidentally* related to this identity.

Additionally, iPSCs open up the possibility of generating gametes in vitro which could then be used to mass-produce human embryos for research and experimentation. The achievement of deriving full-grown mice from ESCs or iPSCs is being used as an argument in favor of such a possible utilization of iPSCs. The argument goes something like this: if one states that the ICM of embryos produced in vitro are to be considered equivalent to the ESCs and iPSCs used to make mice through tetraploid complementation, then embryos before implantation can be considered equivalent to a clump of stem cells.

Recent research also shows that it is possible to directly convert adult somatic cells of a nonneural lineage into neuronal cells without the necessity of going through an undifferentiated stage of iPSCs. One can conjecture that if the technique further improves, the cells obtained this way may supersede iPSCs for use in disease modeling and regenerative medicine, and dream of reprogramming adult cells without having to worry about whether one is making human beings. Nonetheless, direct reprogramming itself also heralds how rapidly these techniques are progressing and that soon it may be possible to obtain iPSCs capable of developing to birth when injected into a blastocyst without ICM.

In light of all these recent achievements, the question of whether or not a cell entity is a human being reaches again an unexpected importance even in the context of reprogramming adult cells.

Alas, this is not the end of the story. Ethical analysis of recent proposals to develop human–animal hybrids for research will also require us to make scientific-philosophical distinctions regarding a certain cell entity's moral status. For if one wishes to perform experimentation on a hybrid entity, it is necessary to identify, insofar as it is possible, what kind of entity is being experimented on. Answering this question is important for ensuring ethical modes of experimentation on hybrid organisms, but also for ensuring those experiments properly comply with legal requirements for scientific research. Indeed, if one believes that no experimentation should be done on human embryos, it becomes especially important that one be able to distinguish whether or not a hybrid composed of genetic material of a monkey and human being ought to be considered a human or a monkey. Again, we are led to the question: when should a cell entity of human origin be considered a human organism? And even assuming that the production of such hybrids would not entail the destruction of human beings (e.g., if they were brought to gestation and birth), we would need to at least ask the question of whether hybridization is ever ethically objectionable?

This book aims to present contributions exploring the status of the different cell entities produced by the recent technologies and especially embryos in vitro, stem cells, and human–animal hybrids. The work was initiated by a group of researchers convened by the Social Trends Institute to study this topic. While some of the chapters here are papers that were delivered during this consultation sponsored by the Social Trends Institute (STI), others were commissioned and added later.

In Chap. 2, *Boris Greber and Hans Schöler* present a general scientific overview and outlook of the reprogramming of somatic cells to both pluripotent stem cells and cells of a different lineage. This chapter hopes to introduce the reader to the relevant scientific data needed to adequately engage with essays introduced later in the volume.

In Chap. 3, *Maureen Condic* discusses the biological and moral status of the embryo in vitro, offering her own account about the moral and organismic status of pluripotent stem cells derived from embryos, direct reprogramming, and Altered Nuclear Transfer (ANT). Drawing on the importance of *organization* as a principle of life – a view advocated by a number of scientists, philosophers, and theologians– she argues that the recent embryological studies indicate molecular criteria present within the single-celled zygote that allow us to delineate between a single-cell entity with a per se developmental trajectory towards adult human organismic functioning and a cell entity, like those produced from research methods like ANT, which lack such organismic status.

In Chap. 4, *Nicanor Austriaco* presents a proposal for determining the organismic status of parthenotes and complete hydatidiform moles. In so doing, he draws upon two important distinctions: that between an active and passive potentiality and that between a whole and a part. Austriaco argues that those cell entities with an *active* potentiality (i.e., a potential actualized wholly from within the entity itself) to develop into a *whole* tumor ought to be considered nonorganisms, whereas those cell entities with a *passive* potentiality (i.e., a potential actualized from without) to develop into a *partial* tumor ought to be considered disabled human organisms. He also argues that a separated Inner cell mass (ICM) of a blastocyst should be considered equivalent to a clump of isolated pluripotent stem cells.

In Chap. 5, *Joachim Huarte and Antoine Suarez* present criteria for distinguishing between embryos and nonembryos that is grounded in the *proper* biological potential for developing the neural activity responsible for controlling spontaneous motility. They argue that the presence of any cellular or genetic deficiency that directly inhibits the appearance of neural activity (DIANA deficiency) is indicative that said cell entity is not, properly speaking, a human being (i.e., a human animal organism). They further argue that their criteria make better sense of the clinical definition of brain death thereby establishing an important relationship between the identification of the presence of life and death in a cell entity of human origin.

In Chap. 6, *Patrick Lee* tackles the question of how can one determine that a product of a reproductive technique is a human embryo or not. Lee proposes a criterion: does the entity produced have the genetic–epigenetic state, and overall organization, such that it will develop itself to the mature stage of a human organism (an organism with a brain that can provide experience suitable to be the substrate of conceptual thought), provided a suitable environment and nutrition? The chapter briefly defends that criterion and applies it to various types of biological entities.

In Chap. 7, *Pablo Requena* shows that the problem of the epistemological status of bioethics has yet to be resolved. He analyzes the category "person" and the various meanings it has acquired in philosophy over the centuries. He presents the catholic view on this theme and on some recent proposals for obtaining ESCs (like in ANT), relying primarily on the Vatican bioethical instruction *Dignitas Personae*. At the same time, Requena stresses the importance of discovering a valid criterion for distinguishing between an embryo and what is an embryo-like entity also in other contexts. So he argues that many of the products of fertilization that perish are probably not human embryos and therefore are not persons, because these cell entities contain genetic aberrations which are incompatible with life.

In Chap. 8, *Manfred Spieker* discusses whether a human being has a right to life, and analyzes the political, philosophical, and social conflict between two undeniable goods: "freedom of research" and "embryonic protection." The frame of reference for Spieker's discussion is the contemporary debate over embryonic experimentation in Germany. Spieker identifies the relevant historical and legal particularities of the German debate, but also attempts to highlight how the issue has been framed by different parties to that debate. The result is an interesting study in how the particularities of a country's social and political history can help to frame the debate over the moral status of human embryos and the limits of science in interesting and complex ways.

In Chap. 9, *Neville Cobbe* discusses the claim that interspecies mixing undermines the uniqueness of being human and discusses various features of human life that allegedly mark it off from nonhuman life. By surveying a host of literature on the subject, Cobbe's contribution gives the reader a comprehensive survey and introduction to the scientific and ethical discussion surrounding hybridization. In the end, he concludes that there is no "smoking gun" to which we can attribute human uniqueness, but rather that one ought to consider Man's uniqueness in a holistic way by viewing the various dimensions held together in humans. He advocates this view as opposed to the view that there might be only one distinctive feature that sets humans apart from nonhumans.

In Chap. 10, *William Hurlbut* discusses recent proposals for human–animal hybridization/experimentation, reflecting on what kind of moral controversies they are bound to bring into focus. Hurlbut formulates a list of specific public policy proposals for guiding such research as well as articulating some general moral principles from which these proposals derive. In all, his reflection aims to promote a deeper understanding and appreciation of what defines the human creature.

In Chap. 11, *Antoine Suarez* argues that reprogramming adult cells can in principle produce a human being and discusses when this may be the case. He shows that to ultimately settle this question a new experiment is required, which is of scientific relevance as well. He goes on to answer various objections raised to the DIANA criteria (proposed in chapter 5) and shows that according to Aristotelian-Thomistic hylemorphism it is possible to have organisms that, in spite of sharing human origin and features of animal life, are not animated by a spiritual soul. This oddity can be avoided by pairing hylemorphism with the principle that the proper biological potential for spontaneous motility is a necessary and sufficient condition to ascertain animation by a spiritual soul.

To conclude, we want to make it clear that although the nature of the questions we asked our collaborators inevitably involves them in discussing matters of contemporary ethical and political relevance, it is not our intention that this volume be seen as an attempt to give definitive answers to those questions. In fact, the authors are far from reaching a consensus. Instead, our aim is to offer various responses to the philosophical problems encountered by attempting to discern the biological and moral status of cell entities of human origin. Nonetheless, in our view, the arguments provided in each chapter of this book represent a decisive step towards establishing the criteria that allow us to decide whether a particular cell entity is a human being.

Chapter 2
Breakthrough in Stem Cell Research? The Reprogramming of Somatic Cells to Pluripotent Stem Cells: Overview and Outlook

Boris Greber and Hans Schöler

Abstract Human embryonic stem (ES) cells are capable of generating all cell types and tissues of the body. As such they represent an attractive source for therapeutic approaches. However, transplanted cells may be rejected by the immune system. One way to address this problem is to generate patient-specific ES cells. This, however, requires the transformation of the genetic program of somatic cells back to that of an early embryonic state. The field of stem cell research and reprogramming is rapidly evolving. This chapter aims at providing background information to understand some of the most exciting recent developments. Subsequently, the different existing strategies of converting somatic cells into ES-like cells will be reviewed and evaluated.

Keywords Cell explantation • Cell fusion • Cell nuclear transfer • Defined factors • Reprogramming

2.1 Background

2.1.1 Stem Cells and Pluripotency

Stem cells are cells, which on the one hand, can proliferate indefinitely while maintaining their identity (self-renewal), but on the other hand they can also form the various, differentiated cell types of the body (pluripotency). Stem cells differ based on their origins and differentiation potential (Table 2.1). In general, the differentiation potential of these cells decreases with the continuous development

B. Greber • H. Schöler (✉)
Department of Cell and Developmental Biology, Max Planck Institute for Molecular Biomedicine, Röntgenstraße 20, 48149 Münster, Germany
e-mail: office@mpi-muenster.mpg.de

A. Suarez and J. Huarte (eds.), *Is this Cell a Human Being?*,
DOI 10.1007/978-3-642-20772-3_2, © Springer-Verlag Berlin Heidelberg 2011

Table 2.1 Definitions of certain terms

Totipotent	Potential to generate all somatic cells of the fetus as well as extra-embryonic tissues Example: fertilized ovum
Pluripotent	Potential to form all somatic tissues and germ cells Example: ES-cells/Inner cell mass of blastocysts
Multipotent	Potential to form multiple cell types within one germ layer lineage Example: hematopoietic stem cells/adult somatic stem cells
Unipotent	Potential to form only one differentiated cell type Example: spermatogonial stem cells

of the embryo. The cells of the developing organism find themselves in a certain sense on a one-way street; there appears to be no way to return to a state of lesser differentiation. Germline cells, however, represent an anomaly in that they form gametes (egg and sperm cells), from which the totipotent zygote is generated after fertilization. In this regard, the germline cells retain the ability to generate a complete organism.

Because of their potential for self-renewal, stem cells can be isolated and propagated in cell culture, i.e., in vitro. In order to stabilize the undifferentiated state of the cells in the culture, however, one must offer them the appropriate signaling molecules. Ideally, this is a matter of substances which the cells encounter in their natural microenvironment (niche). In this manner, pluripotent cells isolated from the blastocysts of mice were successfully cultivated and propagated. These cultivated cells were designated embryonic stem cells (ES cells). Under the proper conditions, ES cells can be maintained in their undifferentiated state and, in principle, indefinitely propagated. If one reinserts these cells into donor blastocysts and transfers them to the uterus of a pseudopregnant mouse, they will participate in the creation of the various tissues of the murine fetus, giving rise to so-called chimeric animals. A similar method even allows one to generate mice whose tissues are derived completely from the inserted ES cells. These results unequivocally demonstrate the pluripotency of ES cells (Jaenisch and Young 2008). With regard to human ES cells, the most stringent confirmation of pluripotency is the formation in lab mice of benign tumors (teratomas), which are composed of tissues from the three blastodermic layers, endoderm, mesoderm, and ectoderm.

The preservation of pluripotency in vitro documented by these tests, as well as the practically unlimited potential for propagation, makes human ES cells an interesting starting point for possible cell replacement therapies. Many serious illnesses which are a great burden on the public health system are based upon cellular deficiencies (Murry and Keller 2008). One example of the implementation and success of cell replacement therapies is the bone marrow transplant, which has been successfully practiced for some time. This procedure is based on the blood-producing characteristics of somatic stem cells (specifically, hematopoietic stem cells) in adult bone marrow. The potential medical value of a particular type of stem cell is, however, all the greater depending upon its differentiation potential, i.e., its ability to develop into various types of cells. The investigation of the potential of adult, somatic stem cells was stimulated by studies which reported the possibility of differentiation beyond the borders of the germline (Schöler 2004). These findings were surprising because somatic stem cells, due to their comparatively mature origins, should not possess pluripotent characteristics. In fact, there are doubts with regard to these results that need to be seriously considered, i.e., the possibility cannot be eliminated that the findings are based on cultivation artifacts, insufficient data interpretation, or inadequately stringent proof of the differentiation potential. As such, the body of evidence dealing with the plasticity of adult, somatic stem cells is unsatisfactory (Jaenisch and Young 2008). By contrast, the pluripotency of ES cells is clearly established and conforms with their early embryonic origins.

2.1.2 Patient-Specific ES cells

Given that after being reinserted into blastocysts, ES cells take part in the development of all somatic tissues, it should also be possible to generate any arbitrary bodily tissue from them in vitro. In order to have ES cells differentiate in the desired direction in the culture dish, it is best to expose them to the signals that are also operative in vivo, i.e., to imitate these signals. In this case, it is necessary to make use of the established knowledge regarding the embryonic development of mammals – which, incidentally, can be used as a good example of the value of basic research for applied purposes (Murry and Keller 2008). A good example in this regard is the differentiation of human ES cells into insulin-producing pancreatic cells under adherent conditions. In a multistep process that simulated embryonic development, Kroon et al. (2008) differentiated pancreatic cells from human ES cells, and after transplantation into mice, these pancreatic cells formed functional beta cells. After glucose stimulation, human insulin could be detected in the blood of the animals. This example shows quite effectively how advanced experiments to use human ES cells for possible cell replacement therapies already are.

A fundamental problem for cell replacement therapies is the possibility of the rejection of the transplanted cells (in the case of human ES cell therapy, the descendants of the ES cells) by the treated patient's immune system (Schöler 2004). The most elegant way to combat this would be the generation of patient-specific ES cells and their subsequent differentiation into the desired tissue. Because human ES cells can only be extracted from preimplantation embryos, however, this method is not viable. Accordingly, the first experiments were attempted to restore adult somatic cells to a state which corresponds to that of human ES cells. For the sake of understanding the techniques employed in these experiments, it will be useful to next clarify the specific characteristics of ES cells in comparison with differentiated somatic cells.

2.2 Comparison of the Characteristics of ES Cells and Differentiated Somatic Cells

2.2.1 Transcription Factors

In addition to their ability to self-reproduce, ES cells, or alternatively, the cells of the inner cell mass of the blastocyst as their in vivo equivalent, are at any time capable of differentiating into the tissues and organs of the three blastodermic layers. The direction in which these cells differentiate is determined by the signaling molecules present in their immediate vicinity. Under conditions which encourage self-reproduction, the genes whose products can trigger differentiation must be kept inactive. Three genes in particular are responsible for this. The *Oct4*, *Nanog*, and *Sox2* genes are all active specifically in undifferentiated ES cells, and all three code for transcription factors, i.e., for proteins which bind to regulatory DNA sequences and thus activate or repress target genes. The expression of the genes *Oct4*, *Nanog*, and (with some qualifications) *Sox2* is turned off when ES cells differentiate. They are inactive in somatic tissues. Oct4 also plays no role in somatic stem cells (Lengner et al. 2007). If *Oct4* is forcefully expressed in somatic cells of mice, the result is a massive expansion of largely undifferentiated cells and the early death of the animal (Hochedlinger et al. 2005). By contrast, the expression of *Oct4* and both other factors is essential in ES cells. If one inactivates them, the ES cells lose the ability to self-reproduce, and they spontaneously differentiate (Ivanova et al. 2006). These three gene products are also directly associated with the specific qualities of ES cells (self-reproduction and pluripotency). These abilities are, however, not limited to ES cells or the inner cell mass of the blastocyst, rather they can also be conserved at various stages of the germline cycle. Accordingly, the pluripotency factors *Oct4*, *Nanog*, and *Sox2* are active in most stages of this "immortal" line (zygote, blastocyst, epiblast, gamete) (Surani et al. 2007).

One significant increase in knowledge regarding the function of the three transcription factors was brought about by the identification and analysis of the target genes (Boyer et al. 2005). It was discovered that *Oct4*, *Nanog*, and *Sox2* activate their own transcription as well as that of both of the others, creating a self-activating circuit. This explains how the self-reproduction of ES cells, which is marked by the continued expression of *Oct4*, *Nanog*, and *Sox2*, is governed at the molecular level. In addition, the combined action of these three factors governs the activation of a multiplicity of different target genes whose products administer functions specific to ES cells; for example, the generation and processing of extracellular signals that indirectly support self-renewal (Greber et al. 2007). The co-regulation of these genes through these three (and perhaps a very few more) factors allows one to draw the conclusion that the self-replication of ES cells is only possible when the integrity of the nuclear machinery which is itself composed of them is intact. Conversely, these factors could be sufficient to establish an ES cell program: the activity of the central transcription factors would lead to the activation of further ES-cell-specific genes whose products execute their function and, to some extent, activate other genes, etc. *Oct4, Nanog,* and *Sox2* represent the core of the self-renewal machinery in ES cells because they are at the top of this hierarchical network.

Ultimately, the self-reproduction of ES cells requires that the differentiation program be repressed. Cellular differentiation, governed by extracellular signals, is mediated through, among other things, the activation of genes for specific transcription factors. A few such factors have the potential to completely change cellular programs. As such, the expression of the *Cdx2* transcription factor in ES cells is, for example, sufficient to cause them to differentiate into extra-embryonic cells (trophectoderm) (Niwa et al. 2005). Therefore, in order to maintain the undifferentiated state of ES cells, genes which code for the proteins with differentiation potential must be and remain repressed. Data from Boyer et al. (2005) indicates that this task is performed by the three proteins, *Oct4, Nanog*, and *Sox2*. Collectively or individually, they bind to the regulatory sequences of a number of transcription factor genes important for development and keep them inactive. Thus, it appears that only a small number of central factors are necessary in order to establish and maintain the gene expression profile (transcriptome) of ES cells.

2.2.2 Epigenetic Mechanisms

The transition of ES cells from an undifferentiated to a differentiated state is not only regulated by specific transcription factors controlling gene expression, but also by epigenetic mechanisms or processes. Important in this context are chemical modifications of DNA and proteins associated with DNA (e.g., histones), which can facilitate or prevent the transcription apparatus from accessing particular gene segments.

DNA Methylation. Thus, the occurrence of DNA methylation in regulatory gene areas (promoters), for example, generally prevents the transcription/expression of a gene. In other words, it functions repressively. The DNA of somatic or differentiated cells is more strongly methylated than the DNA of the cells of the inner cell mass or, more specifically, that of ES cells. This observation stands in direct relation with the observation that cells which have progressed to a stage of further development have more limited potential than cells which occupy an earlier developmental stage. Genes with a relatively high GC concentration in the promoter region (and therefore with CpG islands which are rarely methylated) are mostly active in ES cells (Mikkelsen et al. 2007)[1]. A large number of these genes remain thus in differentiated cells as well. A series of key genes for embryonic development are, however, deactivated in the course of progressing through

[1]For a definition of the terms 'GC concentration' and 'CpG islands' see the Glossary at the end of this book.

differentiation and development. Prominent examples in this context are the genes for the pluripotency factors *Oct4* and *Nanog*, whose promoter regions are methylated in somatic tissues. DNA methylation is therefore a mechanism which fixes completed steps in the processes of development and differentiation, rendering them irreversible (Bird 2002). Should one wish to "artificially" return somatic cells to a pluripotent state, one must reverse the repressive epigenetic modifications, such as DNA methylation in the *Oct4* promoter.

Histone Acetylation. In addition to DNA methylation, a number of different chemical modifications of DNA packaging proteins have been described, most importantly in this context, modifications of histones. One possibility is their acetylation. The binding of acyl groups at the N-terminus of the histone protein neutralizes its positive charge and loosens the interaction with negatively charged phosphate groups in the DNA. This requires in turn a relaxation of the chromatin structure, whereby the DNA becomes more accessible to the transcription machinery. Histone acetylation is therefore generally correlated with gene activation (Bernstein et al. 2007). ES cells do not necessarily express more genes than differentiated cells, but they must always be in a position to activate various differentiation programs. Therefore, the chromatin in ES cells is in a relatively open, uncondensed form and displays a generally more pronounced histone acetylation. The differentiation of the cells is accompanied by a compacting of the chromatin and a loss of chromatin dynamics (Meshorer and Misteli 2006). In the reprogramming of somatic cells, these special characteristics of the ES cell chromatin must be reproduced.

Histone Methylation. A further class of important epigenetic modifications can be found in the methylation of the lysine residues of histones. In contrast to acetylation, the effect of histone methylation on gene expression is determined by the relative methylation position. The methylation of histone H3 at lysine residue 27 functions repressively and is mediated by the so-called polycomb-protein complex. Genome-wide localization studies have shown that polycomb complexes in ES cells bind to a number of genes which play important roles in developmental processes, among others, cellular differentiation (Boyer et al. 2006; Lee et al. 2006). The methylation of histone H3 at lysine residue 4, however, correlates with gene activation (Guenther et al. 2007). One might think that the simultaneous occurrence of activating and repressive histone methylations at particular genes would be mutually exclusive. It was therefore completely surprising to realize that both methylation varieties can be found in the case of some of the genes in ES cells, particularly those genes which code for differentiation factors (Mikkelsen et al. 2007). These "bivalent" genes must be repressed in self-reproducing ES cells, but they are used in the later cellular differentiation which can be induced in the cell at any time. Genes relevant to development can therefore be held in a state of suspension between repression and potential activation (Jaenisch and Young 2008; Bernstein et al. 2007). This hypothesis is attractive, but also disputed, among other things for the reason that bivalent domains do not appear to exist exclusively in ES cells (Mikkelsen et al. 2007). It is, however, clear that a specific pattern of activating and repressive histone methylations can be found in ES cells. And this pattern must also be reproduced in the process of reprogramming somatic cells.

2.3 Approaches to the Reprogramming of Somatic Cells

2.3.1 Reprogramming Through Nuclear Transfer

The explanations earlier should make it clear that returning somatic cells to an early embryonic state (reprogramming) would be an extremely complex process requiring numerous steps that would not be taken as a whole in normal development. Only individual instances of the necessary reprogramming steps are to be found at specific points in the development of a mammal. One example is the widespread demethylation of the DNA of the male pronucleus immediately after the fertilization of the egg cell (Surani et al. 2007). This demethylation/reprogramming activity affects even nuclei that have been transplanted from differentiated somatic cells into unfertilized egg cells from which the nuclei have been removed. The cloned sheep Dolly was living proof that such reprogramming processes can be found in the egg cells of mammals and that a reversal of the development undergone by somatic cells on the molecular level is possible. Although technically challenging, reprogramming through nuclear transfer is conceptually simple: the nucleus/chromosomes are removed from an unfertilized egg cell and the nucleus of a somatic cell is inserted. Because this is not a form of fertilization, the reconstructed egg cell must afterwards be encouraged to divide by means of special chemicals. Ideally, the egg cell then completes in the cell culture the same development into a blastocyst as a normally fertilized egg. In the case of the reproductive cloning of mammals, the embryo is inserted into the uterus of an animal experiencing pseudopregnancy while in the blastocyte stage. By contrast, therapeutic cloning makes use of human blastocysts from this process for the production of patient-specific ES cells (Fig. 2.1).

Reproductive Cloning. The reproductive cloning of mammals is very inefficient, and the few living animals born exhibit anomalies (Hochedlinger and Jaenisch 2006). This is in most cases due to a faulty reactivation of the embryonic developmental program at the molecular level and therefore an incomplete reprogramming of the genome in the transplanted nucleus from the somatic donor cell (Boiani et al. 2002). This is not surprising given that this process is possible for the egg cell for only a short time. Accordingly, the question suggests itself as to whether donor cells which are closer to the natural totipotent zygote with regards to its epigenetic stage of development might be easier to reprogram. Indeed, the efficiency of cloning is higher after the transplant of nuclei from ES cells than, for example, after the transplant of nuclei from embryonic fibroblasts and considerably higher than when using a terminally differentiated cell as the nucleus donor (Hochedlinger and Jaenisch 2006). Still, it remains unclear which factors are responsible for the reprogramming activity in the mammalian egg cell, but it is probable, however, that among them are factors for chromatin remodeling (Yamanaka 2007). Recently, it was shown that not only unfertilized egg cells but also zygotes are capable of reprogramming somatic nuclei – but only when they are artificially arrested in a particular stage of nuclear division (Egli et al. 2007). In this stage, cytoplasmic

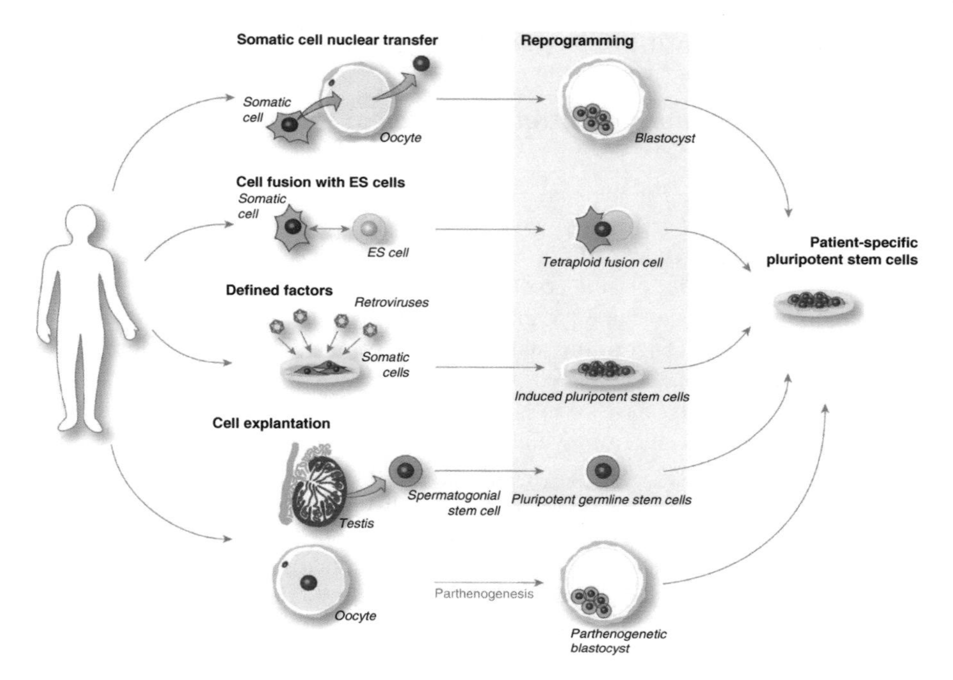

Fig. 2.1 Different approaches to generate patient-specific pluripotent stem cells. The starting material is somatic donor cells in all cases. Reprogramming via nuclear transfer requires donor oocytes in addition. ES cells may be derived from the forming blastocysts. The fusion of somatic cells with ES cells results in ES-like cells with a double set of chromosomes. Direct reprogramming using defined factors is usually based on delivering these by means of retroviruses. Explanted spermatogonial stem cells may – under suitable conditions – give rise to ESC-like cells by spontaneous reprogramming. The generation of parthenogenetic ES cells does not constitute a reprogramming event

reprogramming factors are on hand which are able to affect the inserted somatic chromatin. These results are meaningful in the context of therapeutic cloning (see later) because, as a result of artificial fertilizations, there are more surplus human zygotes than unfertilized eggs frozen and stored worldwide.

Therapeutic Cloning. The application of therapeutic cloning has already been tested in mice. Rudolph Jaenisch and his team were able to successfully correct a gene defect in the animals using the following method (Rideout et al. 2002): they removed the nuclei from somatic cells of diseased animals (immune deficient *Rag2* (−/−)-mice), placed them in enucleated egg cells, and allowed the cells to develop into blastocysts in a cell culture. ES cells were obtained from the inner cell mass of these blastocysts, their gene defects corrected through homologous recombination, and the resulting corrected cells were differentiated into hemapoietic precursors, which were then used in cell therapy in the diseased animals. Advances have also been made in therapeutic cloning in humans and primates. Recently, human blastocysts were generated after cell nuclei from adult connective tissue were inserted into enucleated egg cells (French et al. 2008). In rhesus monkeys, ES cell lines have been produced from cloned blastocysts (Byrne et al. 2007). Interestingly, the ES cells produced through cloning (NT–ES cells) are indistinguishable from those produced conventionally (Brambrink et al. 2006). Regarding their use as a cell therapy, NT–ES cells would therefore have no drawbacks while at the same time having the advantage that they could be produced for a specific patient and would be compatible with the patient's immune system. The fact of the functional equivalence between NT and normal ES cells also suggests that in most cases incomplete reprogramming by the factors in the enucleated egg cell would be completed over the course of the cultivation of the ES cells. ES cells are to a certain extent self-selecting based on functional integrity. This should also mean that ES cells themselves are capable of reprogramming activity.

2.3.2 Reprogramming Through Cell Fusion with ES Cells

That ES cells are capable of reprogramming activity is demonstrated by the fact that the products of fusion between ES cells and somatic cells from mice behave as ES cells. This kind of fusion is induced by electrical or chemical methods after mixing the two cell types (Fig. 2.1). The yield of fused cells is low, but they can be isolated based on previously inserted selection or fluorescence markers. Fused cells demonstrate the morphology of ES cells, are capable of self-reproduction and, to a limited extent, also of potential differentiation in vivo, i.e., after injection into mouse blastocysts. Their participation in the formation of tissues and organs, however, is much lower than normal ES cells (Ying et al. 2002). The explanation for this lies in the tetraploidy of the fused cells. In contrast to diploid ES cells, they contain double the number of chromosomes, which is disadvantageous. That the somatic genome is actually reprogrammed after fusion is demonstrated by epigenetic changes, for example the DNA demethylation in the *Oct4* promoter (Do et al. 2006). The reactivation of the *Oct4* gene in the fused cells occurs surprisingly quickly (Han et al. 2008), indicating the presence of reprogramming factors in the ES cells. The genomes of human somatic cells can also be reprogrammed in this way (Cowan et al. 2005). The tetraploidy of the resulting fused cells is, however, a large obstacle for their potential medical use. There are, of course, already approaches to removing individual chromosomes from the cells (Matsumura et al. 2007), but new methods are needed in order to completely rid the cells of the ES cell genome.

The reprogramming activity of ES cells occurs in the nucleus (Do and Schöler 2004). In order to avoid the problem of tetraploidy discussed earlier, temporarily permeabilized somatic cells are treated with ES cell nuclear extracts in order to reprogram their genome (Taranger et al. 2005). The epigenetic changes induced by the application of this method, e.g., the DNA demethylation in the *Oct4* promoter, are incomplete, however, and so far it has been impossible to achieve true pluripotency in the treated cells.

2.3.3 *Reprogramming Through Cell Explantation*

A further approach to cell reprogramming makes use of the fact that germline cells are related to ES cells/cells of the inner cell mass in that they are totipotent/pluripotent, or rather that they serve to preserve toti/pluripotency across generational borders. This similarity reveals itself on the molecular level, e.g., in the expression of the marker gene for toti/pluripotency, the *Oct4* gene. Spermatogonial stem cells in the testes are clearly unipotent in vivo; they are the progenitors of sperm cells (Table 2.1). One can isolate and cultivate the cells from testicles of mice, however, and three groups have observed that in doing so, colonies can form which are similar to ES cells (Fig. 2.1). In addition, these cells can be propagated using the well-known conditions for ES cell culture. They also show an extensive capacity for differentiation, i.e., they take part in the construction of tissues and organs in developing mouse embryos (chimeras) after they are inserted into mouse blastocysts (Kanatsu-Shinohara et al. 2004; Guan et al. 2006; Seandel et al. 2007). Although it seemed likely, the question remained open as to whether it was actually the spermatogonial stem cells that transformed in the cell culture into something akin to ES cells – though this has also now been demonstrated (Ko et al. 2009). What remains unexplained, however, is how this transformation occurs. The process does not appear to be highly complex given that it occurs spontaneously in the cell culture. In this regard, there are interesting speculations about pluripotency perhaps being the default condition of a cell (Silva and Smith 2008). Up to this point, attempts to establish comparable cells from human testicular biopsies in cell cultures have not succeeded, and experts have cast doubt on publications which suggest doing so (Conrad et al. 2008). The therapeutic potential of such cells is obvious, however.

Autologous ES cells can also be generated from female germline cells and their products, egg cells (oocytes), and even by means other than reprogramming. Oocytes which have not yet completed the second meiotic nuclear division, and are thus diploid, can be encouraged to divide by the use of certain chemicals (parthenogenesis). ES cells can be isolated and cultivated from the resulting blastocysts (Fig. 2.1). Due to epigenetic factors, the developmental potential for such parthenogenetic ES cells must be limited in comparison to normal ES cells. There are studies, however, which suggest equality between the two (Chen et al. 2009). Human ES cells have also already been successfully obtained using this method (Kim et al. 2007).

2.3.4 *Induced Reprogramming: Reprogramming by Means of the Expression of Defined Factors*

Yamanaka's Cocktail. In 2006, Shinya Yamanaka's team reached a milestone in the field of induced reprogramming (Takahashi and Yamanaka 2006). Their work was based on the hypothesis that selected ES-cell-specific factors must be responsible for the reprogramming activity inherent in these cells. Yamanaka and his colleagues selected 24 genes which are important for the preservation of pluripotency in ES cells and expressed them in murine embryonic connective tissue cells. A selection system introduced to these somatic cells that is based on an independent, ES-cell-specific gene (*Fbx15*) should make it possible to isolate successfully reprogrammed cells. In this way, it was possible to show that the combined overexpression of four genes (*Oct4, Sox2, Klf4,* and *c-Myc*) in the connective tissue cells was sufficient to generate cells which resembled ES cells (Fig. 2.1). This technique requires considerable time (ca. 3 weeks) and is quite inefficient (ca. one in ten thousand cells is reprogrammed). The resulting cells are, however, capable of self-reproduction and after injection into mouse blastocysts, appear to take part in the formation of various organs and tissues in the developing mouse embryos. They were therefore designated as "induced pluripotent stem cells" (iPS cells). However, these iPS cells (selected by means of *Fbx15*) differ from ES cells, in that they appear to be incompletely reprogrammed. Exchanging the *Fbx15* selection marker for one which is controlled by the promoters for the *Oct4* or *Nanog* genes, for example, led to iPS cells which cannot be distinguished from ES cells (Maherali et al. 2007; Wernig et al. 2008; Okita et al. 2007). These "second generation" iPS cells fulfill all the criteria for complete reprogramming sketched earlier: reactivation of the endogenous *Oct4* and *Nanog* expression after the demethylation of the respective gene promoters, conformity with the gene expression pattern of ES cells, increased histone acetylation, the activating and repressive histone methylation pattern typical of ES cells as well as full development potential. The latter was recently verified quite impressively by means of the method of so-called tetraploid complementation, whereby living mice are generated almost directly from iPS cells (Zhao et al. 2009; Kang et al. 2009).

Improvements I. The prerequisite for the ability of iPS cells to differentiate into all somatic tissues as ES cells appears to be the inactivation after reprogramming of the four factors previously introduced (Brambrink et al. 2008). When using retroviral vectors to transfer the given factors, this happens automatically in the course of reprogramming. The danger exists, however, that the transgenes will be reactivated over the course of the later differentiation of the cells. *c-Myc* is well known as an oncogene, and 20% of the mice derived from iPS cells did develop tumors (Okita et al. 2007). Therefore, the fact that iPS cells can be generated without the participation of the *c-Myc* gene should be viewed as an important advance (Nakagawa et al. 2008; Wernig et al. 2007). A further motivation to effect direct reprogramming with fewer factors is based on the fact that the integration of the introduced factors into the genome could impair the function of the genes already present. In view of potential cell replacement therapy in humans, this risk cannot be taken. Certain cell types, however, can be reprogrammed with fewer than three or four factors. In order to reprogram neuronal stem cells, which belong to the group of multipotent adult stem cells (Table 2.1), for example, the overexpression of one single gene, *Oct4*, is sufficient, presumably because the other three reprogramming genes are already active in the cells (Kim et al. 2009b). In addition, the potential to generate iPS cells without the previous genetic manipulation of the original cell population, i.e., without the use of genetic selection mechanisms, is of equivalent importance (Meissner et al. 2007). In this context, the increase in the reprogramming rate that occurs with the help of chromatin relaxing substances such as valproic acid is quite promising (Huangfu et al. 2008a). The combination of these modifications makes it possible in the meantime to routinely reprogram human cells with fewer than four factors (Nakagawa et al. 2008; Huangfu et al. 2008b; Takahashi et al. 2007).

Potential Applications and Improvements II. Similarly to NT–ES cells (Rideout et al. 2002), iPS cells have been tested in a mouse model (mouse model for sickle-cell anemia) with regard to their therapeutic potential (Hanna et al. 2007). Attempts to reprogram samples from human patients are not far behind (Park et al. 2008). These kinds of patient-specific iPS cells could at one point be the source material for cell replacement therapies (Fig. 2.1). But they are also well suited to approximately model diseases in the culture dish by differentiating the reprogrammed cells into the diseased cell type (Dimos et al. 2008; Soldner et al. 2009). Such disease models introduce new opportunities to better understand diseases at the molecular level and, for example, to test potential drugs on a much larger scale (Di Giorgio et al. 2008). The current state of iPS technology is already sufficient for these types of applications; in other words, there are no longer any obstacles. The outlook regarding the use of iPS cells for the purposes of cell therapy is, however, quite different. To the degree that this is possible, it must be ensured that the cells to be transplanted demonstrate no genomic lesions – brought about, for example, by the integration of the viral reprogramming factors (regardless of the number used). A number of different approaches have already been tested to combat this problem. Among them are the systematic replacement of single transgenes by means of small molecules (Shi et al. 2008; Lyssiotis et al. 2009) or the cutting out of transgenes using genetic tools after successful reprogramming (Soldner et al. 2009; Kaji et al. 2009; Woltjen et al. 2009). It would naturally be more elegant not to need to work with integrating vehicles but rather with transient overexpression in the form of plasmids, RNA, or recombinant proteins, even if they required repeated application. In fact, this approach has already been successful (Kim et al. 2009a; Zhou et al. 2009; Okita et al. 2008; Yu et al. 2009; Stadtfeld et al. 2008b), but so far, the advantage of virus or integration free reprogramming has unfortunately been bought at the price of reduced efficiency and speed as well as incomplete reprogramming and/or high labor requirements in characterizing the lines. These new approaches are therefore not yet appropriate for routine use.

Mechanism still not understood. The achievements described earlier illustrate the swift development of techniques to induce the reprogramming of somatic cells. But the molecular mechanisms which lie at their roots remain only poorly understood. It is well known that all four genes necessary for reprogramming code for transcription factors. The central functions of the *Oct4* and *Sox2* factors in controlling gene expression specific to ES cells were explained earlier; they are both indispensable for reprogramming. By contrast, the roles of *Klf4* and *c-Myc* remain unexplained. It is currently being discussed whether *Klf4* and *c-Myc* perform various accessory functions in the reprogramming process, for example, to produce an opening of the somatic chromatin in order to allow *Oct4* and *Sox2* access to their target genes (Yamanaka 2007). *Klf4* and *c-Myc* can, however, be exchanged for other factors (Yu et al. 2007). In view of the epigenetic peculiarities of ES cells, this is surprising. It is astonishing, in fact, that the remodeling of chromatin apparently does not require highly specialized proteins. It is important to keep in mind, however, that reprogramming induced by these four factors requires much more time than, for example, reprogramming accomplished by cell fusion. It is possible that important components are still missing. As it stands now, the four factor reprogramming method occurs gradually and asynchronously, i.e., a number of intermediate stages are recognizable (Brambrink et al. 2008; Stadtfeld et al. 2008a). The numerous necessary epigenetic changes occur stochastically, in small steps that are established by the continued expression of the four factors introduced into the cells from outside. Only in a late stage (after approximately 2 weeks) are the endogenous (i.e., cellular) *Oct4, Nanog,* and *Sox2* genes activated and the ability to self-reproduce induced so that the externally introduced factors are no longer necessary.

2.4 Outlook

Reprogramming somatic cells back to a state similar to ES cells lends itself to being a foundation for the future development of patient-specific cell therapy practices. The methods for reprogramming presently available and described earlier have various advantages and disadvantages:

Nuclear Transfer. NT–ES cells have a development potential equal to normal ES cells. Generating them, however, is technically demanding and inefficient. Furthermore, because access to human egg cells (or zygotes) is limited, it is unlikely that therapeutic cloning will ever see broad application. The method itself is also the subject of ethical debate because the production of NT–ES cells requires the creation of preimplantation embryos.

Cell Fusion. The practice of cell fusion makes it possible to quickly and easily reprogram somatic cells. Its serious disadvantage, however, lies in the tetraploidy of the resulting cells. Presently, no methods are available to entirely eliminate the ES cell genome from the fusion products, but it is not impossible that successful methods will be developed in the future.

Cell Explantation. Reprogramming through cell explantation is only possible with stem cells of the germline. Ethical concerns cannot be raised regarding this method, but it is, however, difficult and technically demanding. In addition, reprogramming efficiency drops in relation to the increasing age of the germline cell donor. Human ES cell lines have yet to be generated using this method, but efforts to accomplish this are currently underway.

Induced Reprogramming. Currently, the most dynamic area of development is the induced reprogramming of somatic cells through the introduction of selected factors. The method is attractive because it requires relatively simple means and can be practiced on easily obtainable connective tissue cells, e.g., skin cells. The fact that with the current state of the technique only around one in ten thousand cells is completely reprogrammed is immaterial in practical terms because a cell culture usually contains several times this number of cells. A disadvantage of Yamanaka's "classical" method can be found, however, in the integration of the vectors which the factors code for into the genome of the cells. On the one hand, genes which are important for the cell could be deactivated, and on the other hand, there is the danger of the later reactivation of the genes introduced, which in the case of the *c-Myc* gene, can result in the development of tumors. In view of the possible future use of iPS cells in therapy, the presence of such effects is of course totally unacceptable. Since 2006, however, attempts have been made to minimize the genetic changes in the cells. As stated earlier, these attempts include: (1) the reduction of the number of factors with a consequent reduction in reprogramming efficiency, (2) the substitution of genetic factors for small molecules, (3) the cutting out of the integrated transgenes after reprogramming, or (4) all of the approaches to accomplish reprogramming from the outset without viruses or integration. A silver bullet has not yet been found, because generally any given advantage in one approach is bought at the price of a specific disadvantage. Often, the process of reprogramming is made too long through changes and simplifications of the method. This is then frequently no longer practical and offers too much time for the appearance of spontaneous mutations in the cells. Ironically, it then appears sensible to look for genes or small molecules which can accelerate the reprogramming process. As long as their application is not accompanied by changes to the genotypes of the cells to be reprogrammed, the number of factors used is for all intents and purposes unimportant.

In any case, further improvements are to come with regard to factor-induced reprogramming; it is surely the method with the greatest potential. Recent work demonstrated the transcription factor-mediated conversion of mouse fibroblasts into neurons or cardiomyocytes, respectively, bypassing the isolation of an iPS-like intermediate (Vierbuchen et al. 2010; Ieda et al. 2010). With regards to further facilitating reprogramming to iPS cells, Warren et al. recently demonstrated efficient virus and vector-free reprogramming using synthetic modified mRNA molecules (Warren et al. 2010). It remains to be seen which approach – reprogramming through the formation of iPS cells vs. trans-reprogramming – will be superior with regards to generating defined (patient-specific) differentiated cell types of interest for applied purposes.

References

Bernstein BE, Meissner A, Lander ES (2007) The mammalian epigenome. Cell 128(4):669–681

Bird A (2002) DNA methylation patterns and epigenetic memory. Genes Dev 16(1):6–21

Boiani M, Eckardt S, Schöler HR, McLaughlin KJ (2002) Oct4 distribution and level in mouse clones: consequences for pluripotency. Genes Dev 16(10):1209–1219

Boyer LA, Lee TI, Cole MF, Johnstone SE, Levine SS, Zucker JP, Guenther MG, Kumar RM, Murray HL, Jenner RG, Gifford DK, Melton DA, Jaenisch R, Young RA (2005) Core transcriptional regulatory circuitry in human embryonic stem cells. Cell 122(6):947–956

Boyer LA, Plath K, Zeitlinger J, Brambrink T, Medeiros LA, Lee TI, Levine SS, Wernig M, Tajonar A, Ray MK, Bell GW, Otte AP, Vidal M, Gifford DK, Young RA, Jaenisch R (2006) Polycomb complexes repress developmental regulators in murine embryonic stem cells. Nature 441(7091):349–353

Brambrink T, Hochedlinger K, Bell G, Jaenisch R (2006) ES cells derived from cloned and fertilized blastocysts are transcriptionally and functionally indistinguishable. Proc Natl Acad Sci USA 103(4):933–938

Brambrink T, Foreman R, Welstead GG, Lengner CJ, Wernig M, Suh H, Jaenisch R (2008) Sequential expression of pluripotency markers during direct reprogramming of mouse somatic cells. Cell Stem Cell 2(2):151–159

Byrne JA, Pedersen DA, Clepper LL, Nelson M, Sanger WG, Gokhale S, Wolf DP, Mitalipov SM (2007) Producing primate embryonic stem cells by somatic cell nuclear transfer. Nature 450 (7169):497–502

Chen Z, Liu Z, Huang J, Amano T, Li C, Cao S, Wu C, Liu B, Zhou L, Carter MG, Keefe DL, Yang X, Liu L (2009) Birth of parthenote mice directly from parthenogenetic embryonic stem cells. Stem Cells 27(9):2136–2145

Conrad S, Renninger M, Hennenlotter J, Wiesner T, Just L, Bonin M, Aicher W, Buhring HJ, Mattheus U, Mack A, Wagner HJ, Minger S, Matzkies M, Reppel M, Hescheler J, Sievert KD, Stenzl A, Skutella T (2008) Generation of pluripotent stem cells from adult human testis. Nature 456(7220):344–349

Cowan CA, Atienza J, Melton DA, Eggan K (2005) Nuclear reprogramming of somatic cells after fusion with human embryonic stem cells. Science 309(5739):1369–1373

Di Giorgio FP, Boulting GL, Bobrowicz S, Eggan KC (2008) Human embryonic stem cell-derived motor neurons are sensitive to the toxic effect of glial cells carrying an ALS-causing mutation. Cell Stem Cell 3(6):637–648

Dimos JT, Rodolfa KT, Niakan KK, Weisenthal LM, Mitsumoto H, Chung W, Croft GF, Saphier G, Leibel R, Goland R, Wichterle H, Henderson CE, Eggan K (2008) Induced pluripotent stem cells generated from patients with ALS can be differentiated into motor neurons. Science 321(5893):1218–1221

Do JT, Schöler HR (2004) Nuclei of embryonic stem cells reprogram somatic cells. Stem Cells 22 (6):941–949

Do JT, Han DW, Schöler HR (2006) Reprogramming somatic gene activity by fusion with pluripotent cells. Stem Cell Rev Rep 2(4):257–264

Egli D, Rosains J, Birkhoff G, Eggan K (2007) Developmental reprogramming after chromosome transfer into mitotic mouse zygotes. Nature 447(7145):679–685

French AJ, Adams CA, Anderson LS, Kitchen JR, Hughes MR, Wood SH (2008) Development of human cloned blastocysts following somatic cell nuclear transfer with adult fibroblasts. Stem Cells 26(2):485–493

Greber B, Lehrach H, Adjaye J (2007) Fibroblast growth factor 2 modulates transforming growth factor beta signaling in mouse embryonic fibroblasts and human ESCs (hESCs) to support hESC self-renewal. Stem Cells 25(2):455–464

Guan K, Nayernia K, Maier LS, Wagner S, Dressel R, Lee JH, Nolte J, Wolf F, Li M, Engel W, Hasenfuss G (2006) Pluripotency of spermatogonial stem cells from adult mouse testis. Nature 440(7088):1199–1203

Guenther MG, Levine SS, Boyer LA, Jaenisch R, Young RA (2007) A chromatin landmark and transcription initiation at most promoters in human cells. Cell 130(1):77–88

Han DW, Do JT, Gentile L, Stehling M, Lee HT, Schöler HR (2008) Pluripotential reprogramming of the somatic genome in hybrid cells occurs with the first cell cycle. Stem Cells 26(2):445–454

Hanna J, Wernig M, Markoulaki S, Sun CW, Meissner A, Cassady JP, Beard C, Brambrink T, Wu LC, Townes TM, Jaenisch R (2007) Treatment of sickle cell anemia mouse model with iPS cells generated from autologous skin. Science 318(5858):1920–1923

Hochedlinger K, Jaenisch R (2006) Nuclear reprogramming and pluripotency. Nature 441(7097): 1061–1067

Hochedlinger K, Yamada Y, Beard C, Jaenisch R (2005) Ectopic expression of Oct-4 blocks progenitor-cell differentiation and causes dysplasia in epithelial tissues. Cell 121(3):465–477

Huangfu D, Maehr R, Guo W, Eijkelenboom A, Snitow M, Chen AE, Melton DA (2008a) Induction of pluripotent stem cells by defined factors is greatly improved by small-molecule compounds. Nat Biotechnol 26(7):795–797

Huangfu D, Osafune K, Maehr R, Guo W, Eijkelenboom A, Chen S, Muhlestein W, Melton DA (2008b) Induction of pluripotent stem cells from primary human fibroblasts with only Oct4 and Sox2. Nat Biotechnol 26(11):1269–1275

Ieda M, Fu JD, Delgado-Olguin P, Vedantham V, Hayashi Y, Bruneau BG, Srivastava D (2010) Direct reprogramming of fibroblasts into functional cardiomyocytes by defined factors. Cell 142(3):375–386

Ivanova N, Dobrin R, Lu R, Kotenko I, Levorse J, DeCoste C, Schafer X, Lun Y, Lemischka IR (2006) Dissecting self-renewal in stem cells with RNA interference. Nature 442(7102): 533–538

Jaenisch R, Young R (2008) Stem cells, the molecular circuitry of pluripotency and nuclear reprogramming. Cell 132(4):567–582

Kaji K, Norrby K, Paca A, Mileikovsky M, Mohseni P, Woltjen K (2009) Virus-free induction of pluripotency and subsequent excision of reprogramming factors. Nature 458(7239):771–775

Kanatsu-Shinohara M, Inoue K, Lee J, Yoshimoto M, Ogonuki N, Miki H, Baba S, Kato T, Kazuki Y, Toyokuni S, Toyoshima M, Niwa O, Oshimura M, Heike T, Nakahata T, Ishino F, Ogura A, Shinohara T (2004) Generation of pluripotent stem cells from neonatal mouse testis. Cell 119(7):1001–1012

Kang L, Wang J, Zhang Y, Kou Z, Gao S (2009) iPS cells can support full-term development of tetraploid blastocyst-complemented embryos. Cell Stem Cell 5(2):135–138

Kim K, Ng K, Rugg-Gunn PJ, Shieh JH, Kirak O, Jaenisch R, Wakayama T, Moore MA, Pedersen RA, Daley GQ (2007) Recombination signatures distinguish embryonic stem cells derived by parthenogenesis and somatic cell nuclear transfer. Cell Stem Cell 1(3):346–352

Kim D, Kim CH, Moon JI, Chung YG, Chang MY, Han BS, Ko S, Yang E, Cha KY, Lanza R, Kim KS (2009a) Generation of human induced pluripotent stem cells by direct delivery of reprogramming proteins. Cell Stem Cell 4(6):472–476

Kim JB, Sebastiano V, Wu G, Arauzo-Bravo MJ, Sasse P, Gentile L, Ko K, Ruau D, Ehrich M, van den Boom D, Meyer J, Hubner K, Bernemann C, Ortmeier C, Zenke M, Fleischmann BK, Zaehres H, Schöler HR (2009b) Oct4-induced pluripotency in adult neural stem cells. Cell 136 (3):411–419

Ko K, Tapia N, Wu G, Kim JB, Bravo MJ, Sasse P, Glaser T, Ruau D, Han DW, Greber B, Hausdorfer K, Sebastiano V, Stehling M, Fleischmann BK, Brustle O, Zenke M, Schöler HR (2009) Induction of pluripotency in adult unipotent germline stem cells. Cell Stem Cell 5 (1):87–96

Kroon E, Martinson LA, Kadoya K, Bang AG, Kelly OG, Eliazer S, Young H, Richardson M, Smart NG, Cunningham J, Agulnick AD, D'Amour KA, Carpenter MK, Baetge EE (2008) Pancreatic endoderm derived from human embryonic stem cells generates glucose-responsive insulin-secreting cells *in vivo*. Nat Biotechnol 26(4):443–452

Lee TI, Jenner RG, Boyer LA, Guenther MG, Levine SS, Kumar RM, Chevalier B, Johnstone SE, Cole MF, Isono K, Koseki H, Fuchikami T, Abe K, Murray HL, Zucker JP, Yuan B, Bell GW, Herbolsheimer E, Hannett NM, Sun K, Odom DT, Otte AP, Volkert TL, Bartel DP, Melton DA, Gifford DK, Jaenisch R, Young RA (2006) Control of developmental regulators by Polycomb in human embryonic stem cells. Cell 125(2):301–313

Lengner CJ, Camargo FD, Hochedlinger K, Welstead GG, Zaidi S, Gokhale S, Schöler HR, Tomilin A, Jaenisch R (2007) Oct4 expression is not required for mouse somatic stem cell self-renewal. Cell Stem Cell 1(4):403–415

Lyssiotis CA, Foreman RK, Staerk J, Garcia M, Mathur D, Markoulaki S, Hanna J, Lairson LL, Charette BD, Bouchez LC, Bollong M, Kunick C, Brinker A, Cho CY, Schultz PG, Jaenisch R (2009) Reprogramming of murine fibroblasts to induced pluripotent stem cells with chemical complementation of Klf4. Proc Natl Acad Sci USA 106(22):8912–8917

Maherali N, Sridharan R, Xie W, Utikal J, Eminli S, Arnold K, Stadtfeld M, Yachechko R, Tchieu J, Jaenisch R, Plath K, Hochedlinger K (2007) Directly reprogrammed fibroblasts show global epigenetic remodeling and widespread tissue contribution. Cell Stem Cell 1(1):55–70

Matsumura H, Tada M, Otsuji T, Yasuchika K, Nakatsuji N, Surani A, Tada T (2007) Targeted chromosome elimination from ES-somatic hybrid cells. Nat Methods 4(1):23–25

Meissner A, Wernig M, Jaenisch R (2007) Direct reprogramming of genetically unmodified fibroblasts into pluripotent stem cells. Nat Biotechnol 25(10):1177–1181

Meshorer E, Misteli T (2006) Chromatin in pluripotent embryonic stem cells and differentiation. Nat Rev Mol Cell Biol 7(7):540–546

Mikkelsen TS, Ku M, Jaffe DB, Issac B, Lieberman E, Giannoukos G, Alvarez P, Brockman W, Kim TK, Koche RP, Lee W, Mendenhall E, O'Donovan A, Presser A, Russ C, Xie X, Meissner A, Wernig M, Jaenisch R, Nusbaum C, Lander ES, Bernstein BE (2007) Genome-wide maps of chromatin state in pluripotent and lineage-committed cells. Nature 448(7153):553–560

Murry CE, Keller G (2008) Differentiation of embryonic stem cells to clinically relevant populations: lessons from embryonic development. Cell 132(4):661–680

Nakagawa M, Koyanagi M, Tanabe K, Takahashi K, Ichisaka T, Aoi T, Okita K, Mochiduki Y, Takizawa N, Yamanaka S (2008) Generation of induced pluripotent stem cells without Myc from mouse and human fibroblasts. Nat Biotechnol 26(1):101–106

Niwa H, Toyooka Y, Shimosato D, Strumpf D, Takahashi K, Yagi R, Rossant J (2005) Interaction between Oct3/4 and Cdx2 determines trophectoderm differentiation. Cell 123(5):917–929

Okita K, Ichisaka T, Yamanaka S (2007) Generation of germline-competent induced pluripotent stem cells. Nature 448(7151):313–317

Okita K, Nakagawa M, Hyenjong H, Ichisaka T, Yamanaka S (2008) Generation of mouse induced pluripotent stem cells without viral vectors. Science 322(5903):949–953

Park IH, Arora N, Huo H, Maherali N, Ahfeldt T, Shimamura A, Lensch MW, Cowan C, Hochedlinger K, Daley GQ (2008) Disease-specific induced pluripotent stem cells. Cell 134 (5):877–886

Rideout WM 3rd, Hochedlinger K, Kyba M, Daley GQ, Jaenisch R (2002) Correction of a genetic defect by nuclear transplantation and combined cell and gene therapy. Cell 109(1):17–27

Schöler HR (2004) Das Potential von Stammzellen: Eine Bestandsaufnahme. Bundesgesundheitsblatt Gesundheitsforschung Gesundheitsschutz 47(6):565–577

Seandel M, James D, Shmelkov SV, Falciatori I, Kim J, Chavala S, Scherr DS, Zhang F, Torres R, Gale NW, Yancopoulos GD, Murphy A, Valenzuela DM, Hobbs RM, Pandolfi PP, Rafii S (2007) Generation of functional multipotent adult stem cells from GPR125+ germline progenitors. Nature 449(7160):346–350

Shi Y, Desponts C, Do JT, Hahm HS, Schöler HR, Ding S (2008) Induction of pluripotent stem cells from mouse embryonic fibroblasts by Oct4 and Klf4 with small-molecule compounds. Cell Stem Cell 3(5):568–574

Silva J, Smith A (2008) Capturing pluripotency. Cell 132(4):532–536

Soldner F, Hockemeyer D, Beard C, Gao Q, Bell GW, Cook EG, Hargus G, Blak A, Cooper O, Mitalipova M, Isacson O, Jaenisch R (2009) Parkinson's disease patient-derived induced pluripotent stem cells free of viral reprogramming factors. Cell 136(5):964–977

Stadtfeld M, Maherali N, Breault DT, Hochedlinger K (2008a) Defining molecular cornerstones during fibroblast to iPS cell reprogramming in mouse. Cell Stem Cell 2(3):230–240

Stadtfeld M, Nagaya M, Utikal J, Weir G, Hochedlinger K (2008b) Induced pluripotent stem cells generated without viral integration. Science 322(5903):945–949

Surani MA, Hayashi K, Hajkova P (2007) Genetic and epigenetic regulators of pluripotency. Cell 128(4):747–762

Takahashi K, Yamanaka S (2006) Induction of pluripotent stem cells from mouse embryonic and adult fibroblast cultures by defined factors. Cell 126(4):663–676

Takahashi K, Tanabe K, Ohnuki M, Narita M, Ichisaka T, Tomoda K, Yamanaka S (2007) Induction of pluripotent stem cells from adult human fibroblasts by defined factors. Cell 131 (5):861–872

Taranger CK, Noer A, Sorensen AL, Hakelien AM, Boquest AC, Collas P (2005) Induction of dedifferentiation, genomewide transcriptional programming, and epigenetic reprogramming by extracts of carcinoma and embryonic stem cells. Mol Biol Cell 16(12):5719–5735

Vierbuchen T, Ostermeier A, Pang ZP, Kokubu Y, Südhof TC, Wernig M (2010) Direct conversion of fibroblasts to functional neurons by defined factors. Nature 463(7284):1035–1041

Warren L, Manos PD, Ahfeldt T, Loh YH, Li H, Lau F, Ebina W, Mandal PK, Smith ZD, Meissner A, Daley GQ, Brack AS, Collins JJ, Cowan C, Schlaeger TM, Rossi DJ (2010) Highly efficient reprogramming to pluripotency and directed differentiation of human cells with synthetic modified mRNA. Cell Stem Cell 7(5):618–630

Wernig M, Meissner A, Foreman R, Brambrink T, Ku M, Hochedlinger K, Bernstein BE, Jaenisch R (2007) *In vitro* reprogramming of fibroblasts into a pluripotent ES-cell-like state. Nature 448 (7151):318–324

Wernig M, Meissner A, Cassady JP, Jaenisch R (2008) c-Myc is dispensable for direct reprogramming of mouse fibroblasts. Cell Stem Cell 2(1):10–12

Woltjen K, Michael IP, Mohseni P, Desai R, Mileikovsky M, Hamalainen R, Cowling R, Wang W, Liu P, Gertsenstein M, Kaji K, Sung HK, Nagy A (2009) piggyBac transposition reprograms fibroblasts to induced pluripotent stem cells. Nature 458(7239):766–770

Yamanaka S (2007) Strategies and new developments in the generation of patient-specific pluripotent stem cells. Cell Stem Cell 1(1):39–49

Ying QL, Nichols J, Evans EP, Smith AG (2002) Changing potency by spontaneous fusion. Nature 416(6880):545–548

Yu J, Vodyanik MA, Smuga-Otto K, Antosiewicz-Bourget J, Frane JL, Tian S, Nie J, Jonsdottir GA, Ruotti V, Stewart R, Slukvin II, Thomson JA (2007) Induced pluripotent stem cell lines derived from human somatic cells. Science 318(5858):1917–1920

Yu J, Hu K, Smuga-Otto K, Tian S, Stewart R, Slukvin II, Thomson JA (2009) Human induced pluripotent stem cells free of vector and transgene sequences. Science 324(5928):797–801

Zhao XY, Li W, Lv Z, Liu L, Tong M, Hai T, Hao J, Guo CL, Ma QW, Wang L, Zeng F, Zhou Q (2009) iPS cells produce viable mice through tetraploid complementation. Nature 461(7260): 86–90

Zhou H, Wu S, Joo JY, Zhu S, Han DW, Lin T, Trauger S, Bien G, Yao S, Zhu Y, Siuzdak G, Schöler HR, Duan L, Ding S (2009) Generation of induced pluripotent stem cells using recombinant proteins. Cell Stem Cell 4(5):381–384

Chapter 3
Preimplantation Stages of Human Development: The Biological and Moral Status of Early Embryos

Maureen L. Condic

Abstract There is currently no consensus on when human life begins, and consequently, the biological, moral, and legal status of early human embryos is unclear. Here, the biological facts concerning early human development are examined to establish a scientific view of when human life begins. The evidence clearly indicates that a new human organism (i.e., a human being) is formed at the point of sperm–egg fusion. The events occurring during preimplantation development provide further support for the conclusion that the early embryo is an organism. The status of the zygote as a complete organism is contrasted to that of human pluripotent stem cells that are best understood as *parts* of an organism. Two common arguments for assigning human rights to developing human beings are outlined and their logical implications are briefly discussed, in light of the scientific facts regarding human development.

Keywords Blastocyst • Conception • Human rights • Organism • Preimplantation • Zygote

3.1 Introduction

The biological and moral status of early human embryos has become a matter of discussion in recent years, primarily as a consequence of assisted reproduction technology, human embryonic stem cell research, and human cloning. Prior to the development of procedures for the production of human embryos in the laboratory, there was little need to define precisely when human life begins or to determine the moral status of early human embryos. Yet as the number of "excess" embryos produced

---

M.L. Condic
Department of Neurobiology and Pediatrics, University of Utah School of Medicine Eccles Institute of Human Genetics, Bldg. 533, Room 2280, 15 North 2030 East, Salt Lake City, UT 84112–5330, USA
e-mail: mlcondic@neuro.utah.edu

A. Suarez and J. Huarte (eds.), *Is this Cell a Human Being?*,
DOI 10.1007/978-3-642-20772-3_3, © Springer-Verlag Berlin Heidelberg 2011

in fertility clinics grows, and pressure mounts to use these embryos for biomedical research, establishing an empirical view of early human development becomes a matter of increasing urgency.

Currently, there is little consensus among scientists, philosophers, ethicists, and theologians regarding when human life begins. While many assert that life begins at "the moment of conception", precisely when this moment occurs has not been rigorously defined. Indeed, the legislative bodies of different counties have defined the "moment" of conception quite differently. For example, Canada defines a human embryo as "a human organism during the first 56 days of its development following fertilization or creation,"[1] a definition that is similar to that proposed in the United States.[2] In contrast, Germany sets the beginning of life at a point that occurs approximately 24 h after the fusion of sperm and egg, defining a human embryo as: "the human egg cell, fertilized and capable of development, from the time of fusion of the nuclei."[3] The United Kingdom defines a human embryo as "an egg that is in the process of fertilisation or is undergoing any other process capable of resulting in an embryo."[4] Recent statements by bioethicists, politicians, and scientists have suggested that human life commences even later, at the eight-cell stage (approximately 3 days postfertilization) (for example, Peters 2006), at implantation of the embryo into the uterus [5–6 days postfertilization; Agar (2007), Hatch (2002),[5] or at formation of the primitive streak (2 weeks postfertilization).[6]

Some of the difficulty in considering when human life begins is that, on a cellular level, life is a continuum with gametes (i.e., living sperm and egg cells) fusing to form a single cell embryo (the zygote) that progresses through embryonic development, eventually maturing and producing mature gametes that give rise to the next

[1] "An Act Respecting Assisted Human Reproduction and Related Research (Bill C-6)," Canada/ Government, Ottawa (29 March 2004).

[2] "An organism of the species Homo sapiens during the earliest stages of development, from 1 cell up to 8 weeks", Human Chimera Prohibition Act of 2005, S. 659, 109th Cong. (2005).

[3] "Federal Embryo Protection Law" (1990) (Bundesgesetzblatt, Part 1, December 19, 1990, pp. 2746–2748).

[4] Human Fertilisation and Embryology Act 2008, Part 1, (2). Available at: http://www.opsi.gov.uk/ acts/acts2008/ukpga_20080022_en_1.

[5] Nicholas Agar argues that embryos produced in the laboratory do not have rights because they lack a functional relationship with the womb, a sentiment mirrored by Senator Orrin Hatch.

[6] For example, the Guidelines for Conduct of Human Embryonic Stem Cell Research published by the largest scientific society of stem cell researchers in the world (International Society for Stem Cell Research: http://www.isscr.org/guidelines/index.htm (accessed Dec 1, 2008) do not explicitly define when life begins, yet allow research on human embryos up to 14 days postfertilization, or until the appearance of the primitive streak (typically seen at 14 days).

generation. Throughout this process, cells and structures with radically different properties are formed. Is there actually a point in this continual progression where the life of a new individual human being commences, or is this essentially an arbitrary decision? Moreover, if a new human individual is formed by the union of sperm and egg, what value does this life have at various stages of development and maturation?

3.2 How Does Science Distinguish Between Cell Types?

To address the question of when life begins from a scientific perspective, we must first consider when a new cell that is distinct from sperm and egg is formed. The product of sperm–egg fusion is often referred to as a "fertilized egg" [see, for example, a recent review in the journal *Science* that equates the term "zygote" with a "fertilized egg" (Stitzel and Seydoux 2007)]. Clearly, as long as sperm and/or egg persist, there can be no new human life. Once an oocyte has been penetrated by a sperm, is this cell a new kind of human cell, distinct from either gamete, or is it a modified gamete (a "fertilized egg")?

To determine whether sperm–egg fusion produces a new kind of human cell, it is important to understand how biologists distinguish between different cell types. By what scientific criteria can we know whether the cell produced by the union of sperm and egg is a new kind of human cell, distinct from the gametes?

Scientific distinctions between cell types rest on two criteria: cell composition and cell behavior. Two cell types are known to be different because they contain different components; different genes are either active or silent, resulting in different molecules being produced. Biologists frequently classify cells based on the presence or absence of specific "markers", i.e., molecules that are uniquely or largely characteristic of a particular cell type. Often, differences in cell composition result in differences in cell behavior that are driven by the unique combination of molecules present in the cell. For example, brain cells have characteristic functions or "behaviors" required for information processing, and liver cells have entirely different cell behaviors, appropriate to liver function. The unique functions of both brain and liver cells reflect the unique molecules present in each cell type. When cells have distinct composition and/or distinct behavior, they are classed as distinct cell types. These two criteria are used in all fields of biological science and are the basis for *all* scientific distinctions between cell types, as opposed to distinctions reflecting personal, political, or religious convictions.

Based on both composition and behavior, it is unambiguously clear that a distinct (new) cell type comes into existence at the point of sperm–egg fusion, an event that occurs very rapidly following the initial binding of the sperm and egg surface membranes (reviewed in Primakoff and Myles 2007). The scientific evidence in support of this conclusion has recently been discussed in considerable detail (Condic 2008a), and is summarized in Fig. 3.1. Briefly, at the point of sperm–egg fusion, a single cell is generated that contains all the components of

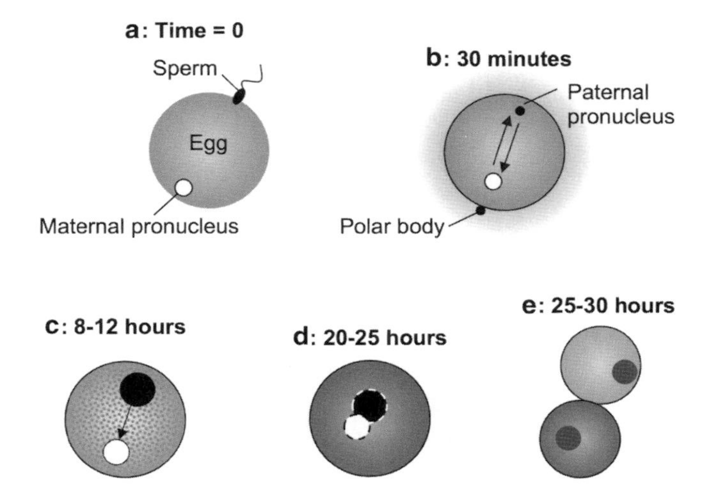

Fig. 3.1 Events occurring in the first day in human development. (**a**) The initial binding of sperm and egg initiates fusion of the two gametes, an event that takes less than a second to complete. Prior to fusion, the maternal nucleus is arrested and has not completed the second meiotic division. (**b**) The zygote forms immediately on sperm–egg fusion. Within the first 1–3 min, changes in calcium within the zygote initiate the cortical reaction, in which chemicals are released that modify the cell surface (*dark line* surrounding the zygote) to prevent additional sperm binding. Factors from the sperm initiate completion of meiosis, and extrusion of the polar body (*arrow* from paternal to maternal pronucleus), to establish the final, diploid genome of the zygote by 30 min. Factors from the egg modify the composition of the both nuclei, but the sperm-derived nucleus is more rapidly and more extensively modified (arrow from maternal to paternal nucleus). (**c**) By approximately 8 h following sperm–egg fusion, the two pronuclei have duplicated their genetic information in preparation for cell division. Zygotic genes become active, producing molecules that alter the composition of the embryo (*gray dots*). Utilization of maternally derived genes is suppressed by factors from the sperm-derived pronucleus. (**d**) By approximately 20 h, the two pronuclei move together and the nuclear membranes break down in preparation for cell division. This event is often referred to as the "fusion" of the two nuclei, or "syngamy." (**e**) Between 25 and 30 h, the zygote divides to produce the two-cell embryo. Evidence suggests that each cell of the two-cell embryo has a distinct developmental bias and that subsequent development requires ongoing cell–cell interactions. *Note*: The zona pellucida, a protective protein coat that surrounds the egg cell (and subsequently the embryo), has not been illustrated, for simplicity

both sperm and egg. This new cell, the zygote or one-cell embryo, is therefore distinct from either sperm or egg in terms of its molecular composition. Thus, based on the first criteria (composition), the zygote is a new cell type.

Following sperm–egg fusion, the zygote immediately enters into a developmental trajectory (i.e., a pattern of cell behavior) that is also distinct from either sperm or egg. The newly formed zygote initiates a sequence of cellular events that block the binding of any additional sperm to the cell surface and sequentially modify the material contributed by sperm and egg to reconfigure it for functions that are unique to the zygote. These actions directly oppose the functions of the gametes that produced the zygote and are part of a zygote-specific developmental sequence that proceeds continuously from the instant of sperm–egg fusion forward. Thus, based on the second criteria (unique behavior), the zygote is also a new cell type, distinct from either sperm or egg.

3.3 How Does Science Distinguish Between Cells and Organisms?

The factual evidence indicates that fusion of sperm and egg immediately and rapidly produces a new human cell, the zygote, in a scientifically well-defined "moment of conception." The zygote is not a modified gamete (i.e., a "fertilized egg") but a unique kind of cell with both molecular composition and developmental trajectory that are distinct from that of an egg cell. Yet is this new human cell a new human *organism*[7] i.e., a new human being? It has been proposed (Condic and Condic 2005) that a human cell and a human organism can be distinguished by using the criteria we have already used to distinguish between different cell types: organisms have a distinctive pattern of behavior and distinct molecular composition, compared to cells with similar origin and/or properties (for example, embryonic stem cells). In particular, the behavior of a human organism (i.e., a human being) is unique and clearly distinguishes it from any type of human cell.

The major feature that distinguishes an organism from a cell is that all parts of an organism act together in a coordinated manner to preserve the life, health, and continued development of the organism as a whole. While individual cells show complex behavior designed to sustain cellular life, they show no higher level of organization or coordinated function. Skin cells removed from the body and maintained in a laboratory will divide to generate more skin cells, ultimately producing a large disorganized mass, but they will not produce multicellular structures or tissues, much less reproduce the body from which they were derived. In contrast, organisms produce cells in a globally coordinated manner to generate an ordered collection of tissues and structures, all of which contribute to the function of the organism as a whole. The coordinated production of diverse, yet functionally integrated, structures is the defining feature of an organism and the basis for distinguishing an organism from a mere living cell.

Are human zygotes human organisms? The zygote clearly exhibits a high degree of coordinated behavior from the moment of sperm–egg fusion onward. Importantly, this behavior is not merely directed toward promoting the life of the zygote as a single cell (as would be the case for a skin cell), but rather it is directed toward the production of distinct cell types that will act in a globally coordinated manner to produce the structures and relationships necessary for the ongoing development of the

[7]Merriam Webster defines "organism" as "(1) a complex structure of interdependent and subordinate elements whose relations and properties are largely determined by their function in the whole and (2) an individual constituted to carry on the activities of life by means of organs separate in function but mutually dependent: a living being." http://www.merriam-webster.com/dictionary/organism (accessed Dec 1, 2008).

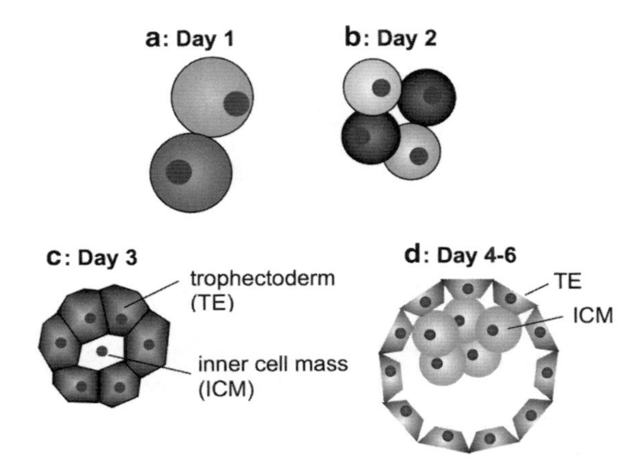

Fig. 3.2 Events occurring between the second and sixth day following sperm–egg fusion. (**a**) At the two-cell stage, cells have a distinct developmental bias and tend to contribute to different tissues in the embryo. (**b**) By the four-cell stage, molecular differences between cells emerge and help to segregate cells into one of the first two cell types in the embryo. By the eight-cell stage, the embryo consists of loosely associated cells and is referred to as a morula. (**c**) Between the 8- and 16-cell stages, the cells of the embryo express new molecules that cause them to tightly adhere and thereby alter the shape of the embryo in a process called compaction. The cells in the outer layer will form trophectoderm (TE) and the inner cells will form the inner cell mass (ICM). Interactions between these two cell types help each to mature in their distinct developmental paths. (**d**) Over the next several days, cells of the embryo continue to divide and specialize. The embryo enlarges to form a hollow ball called the blastocyst. Once the embryo "hatches" from the zona pellucida (not illustrated, for simplicity), the outer TE cells bind to the uterine lining to mediate implantation. TE cells will also generate the placenta and embryonic membranes. ICM cells will generate the majority of the postnatal body

embryo as a whole. The events occurring in the first day of life (Fig. 3.1) anticipate the subsequent events of embryonic development and, indeed, are *required* for subsequent developmental events to occur (Fig. 3.2). Thus, even in the first 24 h of life, there is clear evidence that the zygote behaves as an *organism* (Condic 2008a).

Considering human development over the first week of life greatly strengthens the conclusion that zygote is indeed an organism and not merely a human cell.[8] Fertilization generally takes place in the fallopian tubes, and over the next several days, the embryo slowly travels toward the uterus. Between the first division of the

[8]The great majority of studies done on early mammalian development have been done in mice. Although the general pattern of early development observed in mouse and human embryos is very similar, relatively little is known specifically about the molecular and cellular interactions occurring in early *human* development. However, recent studies of nonhuman primates (reviewed in Byrne et al. 2006) demonstrate a strong degree of similarity between mouse and monkey development, suggesting that much of what has been determined for mouse embryos is likely to be true for human embryos as well.

zygote and implantation into the uterus approximately 5 days later, the embryo continues to act in a coordinated manner to produce the structures and relationships required for the next stages of its own development.

Following the first cell division, utilization (transcription and translation) of zygotic genes increases, and development beyond this stage requires factors produced by the embryo from its own unique genome (Yagi et al. 2007; Nishioka et al. 2008; Hamatani et al. 2004a, b; Worrad et al. 1994; and references therein). Evidence suggests that each cell of the two-cell embryo has a distinct developmental bias and will go on to generate different components of the developing body (Gardner 2001; Fujimori et al. 2003; Piotrowska et al. 2001, Piotrowska-Nitsche et al. 2005). Although cells of early mammalian embryos are highly "plastic" (i.e., able to change their developmental path relatively easily when experimentally perturbed), under normal circumstances, cells appear to enter into distinct and mutually coordinated developmental pathways almost immediately.

By the four-cell stage (Jedrusik et al. 2008; Torres-Padilla et al. 2007), there are clear molecular and developmental differences between cells of the early embryo. Removing cells from the four-cell embryo, or adding extra cells, frequently results in abnormal development or death (Piotrowska-Nitsche et al. 2005), clearly indicating that cells are becoming specialized and producing a mutually coordinated pattern of development. Molecular differences between cells are even more pronounced by the eight-cell stage (Niwa et al. 2005; Hartshorn et al. 2007; Herr et al. 2008; Plusa et al. 2005; Wang et al. 2008; Lu et al. 2008. These early differences are required to establish distinct patterns of gene expression that result in the formation of the first two developmentally unique embryonic tissues (Gardner 2001; Fujimori et al. 2003; Piotrowska et al. 2001, Piotrowska-Nitsche et al. 2005). Thus, in the first three days of life, the embryo has initiated a complex series of molecular events that can only be understood as part of an ongoing, organismal pattern of development.

By 3 days post sperm–egg fusion, at the 8- to 16-cell stage, the embryo undergoes a process called "compaction." New molecules that promote tighter cell–cell attachments are produced, causing the embryo to change its shape to form a solid ball. The changes result in the segregation of cells into the first two distinct tissues of the embryo: the inner cell mass (or ICM) and an outer layer of cells, known as trophectoderm (or TE). Cells of the ICM will mostly contribute to structures of the body that persist after birth, while cells of the TE will produce the placenta and membranes of the embryo.[9]

[9]Structures derived from TE (e.g., placenta) do not contribute to the postnatal body and have classically been considered "extraembryonic" (i.e., outside the embryo), a term that is often misinterpreted as "non-embryonic." However, TE-derived tissues are clearly part of the *embryo*, not part of the mother or of some other entity that coexists with the embryo. Like all embryonic structures, those derived from TE are (1) generated by the embryo, (2) physically continuous with the embryo, (3) genetically identical to the embryo, and (4) critically required for embryonic function and survival. TE-derived structures are best considered transient organs of the embryo and fetus, i.e., bodily structures/organs that function only during prenatal life. Other examples of

Before formation of ICM and TE, the cells of the embryo remain relatively (although not entirely) "plastic." After the ICM and TE form, the differences between these two cell types are stable, with cells rarely converting from one type to another. Thus, by 3 days of age, the embryo has already formed distinct cell types, and ongoing interactions between these cell types are required for the continued health and development of the embryo as a whole.

Following formation of TE and ICM, cells of the embryo continue to multiply and mature. On the 5th or 6th day, the embryo implants into the uterus. For mammals, including humans, establishing contact with the mother is critical for the survival of the embryo.[10] The events of the first five days prepare for this upcoming requirement. On the fourth or fifth day, the embryo undergoes a process known as "hatching," where it is released from a protective protein coat known as the zona pellucida, and the TE is exposed to the uterine environment for the first time. Cells of the TE are uniquely capable of attaching to and interacting with maternal tissues. Therefore, the timing of hatching is critical to ensure that implantation does not occur in an inappropriate environment (the fallopian tube, for example). The coordination of hatching (which is controlled by the embryo itself (O'Sullivan et al. 2004; Perona and Wassarman 1986) with maturation of the TE demonstrates the global coordination of embryonic functions for the sake of the embryo's continued life and health.

Once the embryo has implanted into the uterus, TE cells rapidly form a primitive placenta that begins to function almost immediately, providing nutrients and oxygen to the embryo and modulating the mother's physiology to maintain pregnancy (Licht et al. 2007).

Importantly, both of the earliest cell types of the embryo (TE and ICM) require each other for normal development to proceed. When ICM does not form or is defective, TE cells do not produce a placenta, but a tumor instead (reviewed in Hauzman and Papp 2008; Garner et al. 2007). Similarly, when TE does not form or is defective, ICM cells either die (Chawengsaksophak et al. 1997; (Gardner 2001; Fujimori et al. 2003; Piotrowska et al. 2001, Piotrowska-Nitsche et al. 2005) or do not implant (Meissner and Jaenisch 2006). Thus, the TE and ICM are parts of a coordinated whole and together produce an interactive developmental program that serves the entire embryo. This kind of global interaction of parts for the good of the entity as a whole is the defining characteristic of an organism.

The differentiation of cells into ICM and TE requires complex interactions that begin at the one-cell stage and progress seamlessly throughout the first 5 days of development. Although the early embryo does not look like a fetal or newborn

---

organs with transient functions are the lungs (functioning only during postnatal life) and the ovaries (functioning as reproductive organs only between puberty and menopause).

[10]Unlike birds or amphibians that have large eggs, with enough nutrients to support the developing embryo until hatching, mammals have relatively small, nutrient-poor eggs that can only support development for a few days. Without the mother as a source of food and oxygen, the embryo would not survive. Thus, implantation and establishing contact with the circulatory system of the mother are critical to the health and development of the embryo as a whole.

human being, it exhibits the kind of global coordination of parts for the sake of the whole that is characteristic of an organism. At early stages of development, the cells and tissues produced by the embryo are necessarily very "basic" (in the sense of 'foundational'); they are relatively simple tissues that have not yet diversified into the many cell types and structures characteristic of more mature stages. However, these "basic" tissues are also extraordinarily "potent"; they are capable of generating all of the diverse cells required for development as well as those found in the mature body. Moreover, the developing embryo produces the more familiar structures of the mature body in an orderly and coordinated manner, just as it produces the first two cell types. This unique behavior of the human organism at the zygotic and early embryonic stages provides a clear indication that the cell formed by fusion of sperm and egg is indeed a whole and complete human organism at the earliest stage of human development.

3.4 What Is the Status of Pluripotent Stem Cells Derived From Embryos, Direct Reprogramming and Altered Nuclear Transfer (ANT)

A number of entities are quite similar to human embryos, either in their origin or in their developmental capabilities. Based on the criteria employed here to distinguish human cells from human organisms, we can begin to consider the biological status of entities that may initially appear to be very similar to embryos, such as human parthenotes, hydatidiform moles, clones, and pluripotent stem cells [for discussion of parthenotes, hydatidiform moles, and clones, see Condic (2008a)]. Here, the biological status of pluripotent stem cells is compared to embryos, based on the scientific criteria outlined above.

Pluripotent human stem cells are able to produce all of the cells of the mature human body, and thus they have properties that are very similar to the cells of the ICM in normal embryos. Moreover, under specific conditions in animal models (Nagy et al. 1990; Poueymirou et al. 2007), such stem cells can be induced to form all or the majority of the mature animal's body, raising concern that they are in some sense the functional equivalents of an embryo. Pluripotent stem cells can be obtained from a number of different sources (Fig. 3.3), each of which raises specific ethical (Rao and Condic 2008) and practical (Condic and Rao 2008) concerns. The four major sources of pluripotent stem cells (human embryos, cloned human embryos, altered nuclear transfer, and direct reprogramming) are outlined briefly below.

The most widely studied pluripotent cells are embryonic stem cells (ESCs). Human ESCs are obtained from embryos produced by fertilization, typically "excess" embryos that were generated in the course of assisted reproduction treatments, but not used for reproductive purposes. To produce an ESC "line" (i.e., a stem cell culture derived from a single individual), an embryo is destroyed and its cells dispersed (Fig. 3.3a). The cells originally found in the ICM are harvested and maintained in culture to generate an ESC line.

Pluripotent stem cells can theoretically be derived from cloned human embryos produced by somatic cell nuclear transfer, or SCNT. (Stem cells have been derived from cloned animal embryos, but not yet from cloned humans.) In this procedure, the nuclear material of an egg cell is removed and this empty, "enucleated" egg is fused with a body (i.e., somatic) cell from a mature person. In rare cases, this recombined cell will be a zygote that is genetically identical to the individual who donated the body cell and will begin to develop along a relatively normal pathway. Pluripotent stem cells could be obtained from such cloned human embryos, in the same manner they are currently obtained from embryos produced by fertilization (Fig. 3.3b).

A modification of SCNT has been proposed, in which molecules present in the body cell and/or the egg are manipulated prior to the fusion (i.e., "transfer" of the manipulated body cell nucleus to the enucleated oocyte) such that an embryo does not form, but rather a pluripotent stem cell (Fig. 3.3c) [for a thorough discussion of ANT, see Condic (2008b)]. This procedure, known as altered nuclear transfer, or ANT, has not been done using human cells, but has been successful in generating pluripotent animal stem cells (Meissner and Jaenisch 2006).

Finally, recent work has shown that adult body cells can be directly reprogrammed to a state that is functionally identical to that of a pluripotent stem cell derived from a human embryo, by the addition of a small number of molecules to the adult body cell, in a procedure that does not involve human eggs (Takahashi et al. 2007; Yu et al. 2007). Reprogramming produces induced pluripotent stem cells, or iPSCs, that are genetically identical to the individual donating the original body cell (Fig. 3.3d). Importantly, no egg cells are used and no human embryos produced or destroyed.

Regardless of the source, pluripotent stem cells are distinct from embryos on both of the criteria employed to distinguish human cells from human organisms (composition and behavior). Pluripotent stem cells have patterns of gene expression that are clearly distinct from embryos, utilizing genes that are normally associated with the cells of the ICM and are not active in the zygote or in cells of the TE (Boyer et al. 2005; Adjaye et al. 2005; Babaie et al. 2007; Cauffman et al. 2009). Thus, in terms of their molecular composition, pluripotent stem cells most closely resemble cells of the ICM and are clearly distinct from both zygotes and later stage embryos (consisting of both ICM and TE).

Pluripotent stem cells also show behavior that is not similar to that of an embryo. Stem cells do not act as integrated wholes, but rather as a *part* of a whole. For example, when ESCs are cultured under conditions that promote their differentiation into mature cell types, they are not capable of regenerating the embryo from which they were derived, but rather produce a disorganized mass of tissues and cell types. Such cultures may contain a wide range of different cell types, but the cells and tissue produced have no functional relationship to each other or to anything resembling a whole. Similarly, when pluripotent cells from any source are placed into adult mice, they produce disorganized tumors that contain multiple mature cell

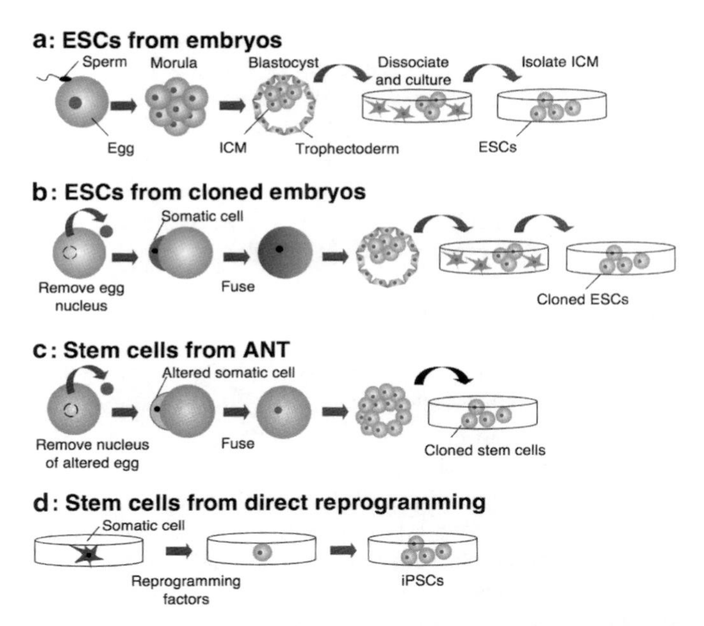

Fig. 3.3 Different ways of obtaining pluripotent stem cells. (**a**) Embryos derived from fertilization are allowed to proceed to the blastocyst stage, when cells differentiate into TE and ICM, then dissociated, and placed in culture. Cells derived from ICM are isolated to establish an embryonic stem cell (ESC) line. (**b**) Somatic cell nuclear transfer (SCNT) involves fusing a mature somatic cell to an enucleated egg. In rare cases, the fused cell is a zygote and can be allowed to develop as in (**a**) and then dissociated to yield embryonic stem cells that are genetically identical to the somatic cell donor. (**c**) Altered nuclear transfer (ANT) is similar to SCNT, but the egg and/or the somatic cell are modified prior to fusion (*gray dots* in egg cell, light gray somatic cell) so that upon fusion, an embryo is not formed; yet pluripotent stem cells that are genetically identical to the somatic cell donor are produced. (**d**) Adult somatic cells can be directly reprogrammed to a pluripotent state (induced pluripotent stem cells, iPSCs) by expressing a small number of reprogramming factors in the somatic cell. iPSCs are genetically identical to the somatic cell donor. [Figure modified from: Mahendra Rao and Maureen L. Condic, "Alternative sources of pluripotent stem cells: scientific solutions to an ethical dilemma", Stem Cells and Development, 17, no. 1, (2008): 1]

types in a chaotic mass. Thus, based on both molecular composition and on cell behavior, pluripotent stem cells are not organisms, but most closely resemble "parts" of an organism, i.e., cells of the ICM that are able to produce all of the cell types found in the mature body, but are incapable of *organizing* these cells into a coherent whole. Thus, pluripotent stem cells do not, by themselves, constitute a whole human being.

Nonetheless, some scientific writers have asserted that embryonic stem cells are the equivalent of human embryos. For example, Dr. Lee Silver states that embryonic stem cells can "produce a human being" (Anderson and Condic 2008). Assertions like this are based on a confusion between what a stem cell can do on its own and what it can be induced to do under special circumstances. When pluripotent stem cells are injected into intact mouse embryos at early stages (morula or blastocyst; see Fig. 3.2c, d), they will contribute randomly to the tissues of the maturing mouse embryo. In this sense, stem cells can become part of an existing embryo. Moreover, if stem cells are injected into an embryo with diminished ability to form ICM (either because it has been experimentally manipulated or because it is at a very early stage of development, before the cells are fully differentiated into TE and ICM), the injected stem cells will form all or most of the ICM and produce the majority of the mature body of the mouse. [This can be done in one of two ways, in a process known as tetraploid complementation or in a slightly more efficient procedure known as laser-assisted injection. See Nagy et al. (1990) and Poueymirou et al. (2007).] Importantly, in both of these situations, the stem cells are not producing the mature body tissues *on their own*; they are responding to an existing embryonic environment and contributing to an already ongoing embryonic pattern of development, in collaboration with the other cells of the embryo. They are acting as a *part* of a mouse embryo, not as a whole organism.

Finally, the prospect of using pluripotent stem cells to produce gametes (egg and sperm), which are subsequently used to make embryos, has led some to claim that there is no substantive difference between stem cells and embryos. If stem cells can produce gametes, and gametes can produce embryos, then embryos would be viewed "like any other type of cell line. They would become objects and would be used as objects" [comments of Davor Solter, reported in Pearson (2008)]. This assertion conflates the ease with which something can be obtained with the nature of the thing itself. If stem cells could be induced to generate gametes, embryos could be produced in very large numbers from these gametes without the difficulties currently associated with obtaining human eggs from women donors. Yet, there would still be a significant difference between a gamete and an embryo; gametes are human cells, but embryos are entire, living human beings. The rather horrifying prospect that embryos could be produced in industrial quantities would in no way alter the clear difference between a cell (e.g., a stem cell or a gamete) and an organism.

3.5 Are Embryos Human Beings with Rights?

Based on the scientific evidence, the zygote formed by fusion of sperm and egg is a human organism, i.e., a human being at the earliest stage of development. Despite the small size and unfamiliar form of the zygote, this single cell is indeed a complete and living organism and member of the human species. Yet in the face of this incontrovertible scientific conclusion, the intuitions of many people (including many scientists) simply stage a revolt; it seems so preposterous to call a zygote – a single cell that is smaller than a pin head – a "human being" that many simply reject this conclusion as too outrageous to be true. Attempting to reconcile this strong intuitive reaction with the undisputed scientific facts, many people turn to the concept of "value," asserting that while embryos are human organisms in a technical sense, they only gradually accrue "human rights" as they mature. Rather than assigning human rights to embryos at all stages of development based on their status as human organisms, rights are linked to something else that is deemed more intuitively reasonable. In this view, while the human organism formed at sperm–egg fusion has the "potential" to become a human person, is not an "actual" person until some later time.

There are two common forms of the argument that "value" and human rights accrue gradually over development: the "structure/function" and the "social convention" forms. The logical implications of each argument are briefly considered, in light of the scientific evidence.

3.5.1 Structure/Function Form: Human Rights Depend on Attainment of Specific Structural and/or Functional Landmarks that are Characteristic of Mature Humans

As discussed above, once a human organism comes to be, development is a seamless process that gradually produces increasingly complex structures and relationships. Eventually, human embryos acquire a more familiar form, coming to "look like" human babies (albeit very small babies, only 2–3 cm in length) around the 7th or 8th week of life. As the structures of the embryo (and later, the fetus) mature, they not only begin to look like the structures we see in a newborn, they also begin to function in ways that are similar to the functions they will have in postnatal life. The point at which a developing organ assumes a "mature" function depends somewhat on the organ and what role it plays in human physiology. The heart, for example, is critical for survival of the embryo even at very early stages, and correspondingly, it begins to function as a primitive blood-circulating organ as early as 20 days post sperm–egg fusion. In contrast, the lungs, which are not required at all during prenatal stages, mature relatively late in development and only begin to function following birth.

In an attempt to link human rights/value to a biological trait other than status as a human organism, many turn to the development of structures and functions seen as "characteristically human", most typically, maturation of the nervous system or attainment of "viability."

Of these two, viability is by far a weaker argument. The ability of a fetus to survive following preterm birth reflects a number of factors, most critically, the availability of sophisticated neonatal intensive care. The lower limit of fetal viability has shifted to progressively earlier developmental stages over the last 10 years, as such technology has become more widely available and now hovers around 22–23 weeks of gestation in most medical centers (Morgan et al. 2008). As technology continues to advance, even younger fetuses are likely to be considered "viable." Consequently, linking human rights to viability provides an almost purely technological definition of who is and who is not the subject of human rights.

In the interest of identifying a *biological*, rather than a *technological* definition of human rights, most proponents of the "structure/function" view reject "viability" as the standard and instead advocate a criterion based on brain maturation. Mental functions such as memory, language, and reasoning produce some of our most intimate and unique experiences as humans, and thus the capacity of the brain to function in these characteristically human ways seems to many to be an appropriate basis for defining who is a subject of rights.

The point at which the nervous system is considered mature enough to warrant conferring rights on a developing human being varies considerably, from the onset of neural function at approximately 10 weeks (Himma 2003), to the onset of "coordinated neural function" at approximately 23 weeks (Penner and Hull 2008), to the establishment of more complex brain connections that are capable of supporting "consciousness," between 30 and 35 weeks (Burgess and Tawia 1996). The variation in these figures illustrates the first serious problem with linking human rights to brain function. In contrast to the fundamental changes that occur at the initial formation of the zygote, maturation of the brain is a continuous process with no clear point at which "landmarks" in brain function are achieved. Indeed, some have argued that the brain is insufficiently mature at birth to warrant calling a newborn a person (Kuhse and Singer 1988). Thus, using the development of the brain as a basis for assigning human rights is necessarily arbitrary and leaves the criteria for determining who is the subject of such rights ill defined.

Nonetheless, many real distinctions are similarly arbitrary and ill-defined, for example, the difference between an "adolescent" and a "child." While everyone agrees there are real and significant differences between these two stages of life, precisely when this transition occurs cannot be defined with accuracy. Although there are markers of puberty, the timing of these events varies considerably and the transition is gradual, rather than abrupt. Just so, the proponents of the structure/function definition claim, with the acquisition of human rights.

Yet this analogy fails in a fundamental way. Adolescence is a stage of physical maturation that is gradually achieved. In contrast, "human being" is something you either are, or are not. While the timing of physical maturation is variable, puberty does not transform an individual from one kind of thing into another. A child does not mature into a horse or an oak tree, but rather continues along the developmental trajectory he or she has followed since the fusion of sperm and egg produced the unique human organism he or she is: that of a *human being*. Thus, the second, and even more serious problem with linking the acquisition of rights to brain maturation (or to the maturation of any biological structure), is that it asserts that a fundamental change has occurred, from a nonperson (or collection of human cells) to a person with rights and value, without *any change in the biological nature of the entity under consideration*. The developmental trajectory and essential organization of a fetus do not change at 10, 20, or 30 weeks of gestation. Maturation in the biological state of an organism simply does not provide any justification for asserting a fundamental change – one that converts a nonperson to a person – has occurred.

Finally, linking human rights to brain function raises the real and profound issue of how we are to view the wide range of variation in brain function among postnatal humans. Are we to assign human rights proportionate to brain function, with the intelligent enjoying greater freedom and privilege than the less intelligent? Most of us find this option intuitively repugnant, and yet it is entirely consistent with the logic of assigning rights to fetuses based on maturing brain function.

Alternatively, are we to identify a minimal level of brain function that qualifies an individual for human rights, and if so, what might this level be? Surely, the very immature level of brain function seen in a 30–35 week fetus does not qualify as a distinctively "human" level of cognitive ability. Setting the minimum level this low would seem to undermine the very premise of the structure/function argument, by linking human rights to a level of cognitive ability that is clearly shared by many nonhuman animals. Indeed, due to the extremely long period of time over which the human brain develops [the human brain continues developing for a very long period following birth and does not achieve characteristically mature structure or function until 20–25 years of age; see, for example, Gogtay et al. (2004) and Sowell et al. (2003)], setting the onset of human worth at *any* embryonic, fetal, or even newborn stage simply cannot be justified based on an argument that human rights depend on acquisition of distinctively human cognitive abilities.[11]

If we are to set a higher, more characteristically human standard of brain function as the basis for assigning rights, how are we to consider those who fall below this minimum level as a consequence of injury, disease, or old age? What of individuals who, due to a developmental or genetic defect, never achieve the minimum requirements to qualify as the subject of rights? Are such individuals to be considered a kind of subhuman animal, to be destroyed at will, or to be used for labor, scientific research, or other purposes? Again, most find this suggestion repugnant and absurd, insisting that mentally impaired individuals nonetheless have human rights. Indeed, more than 130 countries have recently signed a United Nations Convention that asserts mentally impaired persons have an inherent right to life that is on an equal basis with the non-impaired and that further requires countries to prohibit discrimination and guarantee equal legal protection for such persons.[12] Yet if brain function is insufficient to confer human rights (and, as we have seen, "viability" is a technological, rather than a biological criterion) what *other* 'characteristic human structure or function' is sufficient to define those with impaired cognitive ability as persons while simultaneously denying this status to human embryos and human fetuses?

---

[11]It is sometimes claimed that once the brain has acquired sufficient maturity to "potentially" support cognition, actual manifestation of cognitive acts is not required for an individual to be considered human, but this argument does not define "potential" in terms of either biological structures or neurological function and thus seems to be a version of the "Social convention form" discussed below.

[12]The United Nations Convention on the Rights of Persons with Disabilities and Optional Protocol (http://www.un.org/esa/socdev/enable/dispaperdes0.htm; accessed Dec 1, 2008). Entered into force on 3 May 2008, after receiving its 20th Ratification. A total of 67 countries have ratified the Convention or Optional Protocol.

3.5.2 Social Convention Form: Rights Are Assigned Based on Social Constructs that Have No Strict Definition, But Correlate with a "Common-Sense" Definition of What it Means to be Human

The second common form of the argument that rights and value accrue gradually over developmental time is that rights are acquired not as a function of biological state, but entirely as a matter of social convention. This form of the argument abandons all pretense of linking rights to any feature of the developing human being and starts with the assumption that human rights are assigned purely on the basis of pragmatic social concerns (see, for example Rubenfeld 1991). From the perspective of biology, this view is more consistent with the continuous nature of development and the lack of any fundamental change in the status of a developing human organism across its life span.

The appeal of this argument is its apparent common-sense reasonableness. As a society, we grant and revoke rights all the time. It seems natural that children acquire the right to drink and to vote only as they attain a certain level of maturity. Conversely, we are comfortable revoking some rights, for example, the right to drive, as people age and become physically incompetent to control a car. Thus, the right to drive is both gradually acquired and gradually lost, entirely on the basis of pragmatic considerations and social convention. So too, proponents of the social convention view argue, an embryo or a fetus gradually acquires rights as it matures, including the right to continued existence, entirely as a matter of pragmatic social conventions. In this view, there is no "fundamental" right to anything, but merely judgments regarding the appropriateness of conferring a right on an individual, based on their biological state at the time and on the concerns of other interested parties. This view does not deny that an embryo or fetus is a human being, but asserts that society has the power to decide whether human beings have rights or not, including the right to continue living.

The difficulty with this position is that in rejecting the notion that some rights are inalienable (such as the right to life, liberty and the pursuit of happiness) and considering all rights to be matters of social convention, the very notion of "rights" is fatally undermined. This view reduces the notion of "rights" to matter of social accord or contract – agreements that can be manipulated (or eliminated) by those with sufficient power to impose their own preferences. Indeed, with no value or right adhering to persons intrinsically, the concept of "person" becomes entirely meaningless. How is society to determine what rights are "appropriate" to any human organism (person or not), if all rights are arbitrarily assigned? Why should anyone have "rights" beyond those they are able to acquire by force, purchase, or persuasion? What basis is there for treating the weak, the poor, or the less intelligent as equivalent to the strong, the rich, and the clever? To what do we pin concepts of justice and fairness, once human beings enjoy rights merely as a matter of social convention? If society chooses to deny rights to those of a particular race, religion, or gender, on what basis could proponents of the 'social convention' view possibly object? Despite its apparent reasonableness, the notion that rights are nothing more than a matter of public consensus, to be conferred on embryos and fetuses at the discretion of society, undoes the very notion of rights for *all* members of society, regardless of their stage of biological maturation.

3.6 Conclusion

The life of a human being (i.e., a human organism) begins at a scientifically well-defined moment: the fusion of sperm and egg. Sperm–egg fusion produces a one-cell embryo, the zygote, that is distinct from the gametes in terms of both its molecular composition and its pattern of development. From the instant of sperm–egg fusion onward, the embryo acts in a coordinated, organismal manner to produce and to regulate its own development. All of the actions of the embryo are directed toward producing the structures and relationships required for the ongoing life and health of the embryo as a whole. At no time does the embryo even remotely resemble a mere human cell or collection of human cells. Despite the claims of some scientists, pluripotent stem cells are not the equivalents of embryos and they cannot produce an entire living organism on their own. Although pluripotent stem cells can produce many different types of mature cells, and are therefore similar to embryos in some respects, they do not exhibit the global coordination of parts for the sake of the whole that is the hallmark of a living organism. The moral and ethical status of early human embryos cannot be directly addressed by the scientific evidence. However, the arguments that human rights accrue gradually over developmental time are either logically or scientifically flawed and thus do not provide a cogent argument for defining when a human embryo or fetus is to be considered the subject of human rights and worthy of protection.

References

Adjaye J et al (2005) Primary differentiation in the human blastocyst: comparative molecular portraits of inner cell mass and trophectoderm cells. Stem Cells 23(10):1514

Agar N (2007) Embryonic potential and stem cells. Bioethics 21(4):198

Anderson R, Condic ML (2008) Professor Lee Silver's vast scientific conspiracy. *First Things*, http://www.firstthings.com/onthesquare/? p=946. Accessed 1 Dec 2008

Babaie Y et al (2007) Analysis of Oct4-dependent transcriptional networks regulating self-renewal and pluripotency in human embryonic. Stem Cells 25(2):500

Boyer LA et al (2005) Core transcriptional regulatory circuitry in human embryonic stem cells. Cell 122(6):947

Burgess JA, Tawia SA (1996) When did you first begin to feel it? Locating the beginning of human consciousness. Bioethics 10(1):1–26

Byrne JA, Mitalipov SM, Wolf DP (2006) Current progress with primate embryonic stem cells. Curr Stem Cell Res Ther 1(2):127

Cauffman G et al (2009) Markers that define stemness in ESC are unable to identify the totipotent cells in human preimplantation embryos. Hum Reprod 24(1):63
Chawengsaksophak K et al (1997) Homeosis and intestinal tumours in Cdx2 mutant mice. Nature 386:84
Condic ML (2008) When does human life begin? A scientific perspective. Westchester Institute White Paper 1, no. 1: 1. Available at: www.westchesterinstitute.net. Accessed Dec 1, 2008
Condic ML (2008b) Alternative sources of pluripotent stem cells: altered nuclear transfer. Cell Prolif 41(1):7
Condic ML, Condic SB (2005) Defining organisms by organization. Natl Cathol Bioeth Q 5:331
Condic ML, Rao M (2008) Regulatory issues for personalized pluripotent cells. Stem Cells 26 (11):2753
Fujimori T et al (2003) Analysis of cell lineage in two- and four-cell mouse embryos. Development 130 (21):5113
Gardner RL (2001) Specification of embryonic axes begins before cleavage in normal mouse development. Development 128:839
Garner E et al (2007) Gestational trophoblastic disease. Clin Obstet Gynecol 50(1):112
Gogtay N et al (2004) Dynamic mapping of human cortical development during childhood through early adulthood. Proc Natl Acad Sci USA 101(21):8174
Hamatani T et al (2004a) Global gene expression analysis identifies molecular pathways distinguishing blastocyst dormancy and activation. Proc Natl Acad Sci USA 101(28):10326
Hamatani T et al (2004b) Dynamics of global gene expression changes during mouse preimplantation development. Dev Cell 6(1):117
Hartshorn C et al (2007) Single-cell duplex RT-LATE-PCR reveals Oct4 and Xist RNA gradients in 8-cell embryos. BMC Biotechnol 7:87
Hatch O (2002) Hatch makes the case for Regenerative Medicine. Press release, 30 April 2002. http://hatch.senate.gov/newsite/index.cfm?FuseAction=PressReleases.Detail&PressRelease_id=182533&Month=4&Year=2002. Accessed 6 June 2011
Hauzman EE, Papp Z (2008) Conception without the development of a human being. J Perinat Med 36(2):175
Herr JC et al (2008) Distribution of RNA binding protein MOEP19 in the oocyte cortex and early embryo indicates pre-patterning related to blastomere polarity and trophectoderm specification. Dev Biol 314(2):300
Himma KE (2003) What philosophy of mind can tell us about the morality of abortion: personhood, materialism, and the existence of self. Int J Appl Philos 17(1):89
Jedrusik A et al (2008) Role of Cdx2 and cell polarity in cell allocation and specification of trophectoderm and inner cell mass in the mouse embryo. Genes Dev 22(19):2692
Kuhse H, Singer P (1988) Should the baby live?: the problem of handicapped infants. Oxford University Press, Oxford
Licht P et al (2007) Is human chorionic gonadotropin directly involved in the regulation of human implantation? Mol Cell Endocrinol 269(1–2):85
Lu C-W et al (2008) Ras-MAPK signaling promotes trophectoderm formation from embryonic stem cells and mouse embryos. Nat Genet 40(7):921
Meissner A, Jaenisch R (2006) Generation of nuclear transfer-derived pluripotent ES cells from cloned Cdx2-deficient blastocysts. Nature 439(7073):212
Morgan MA et al (2008) Obstetrician-gynecologists' practices regarding preterm birth at the limit of viability. J Matern-Fetal Neonatal Med 21(2):115
Nagy A et al (1990) Embryonic stem cells alone are able to support fetal development in the mouse. Development 110(3):815
Nishioka N et al (2008) Tead4 is required for specification of trophectoderm in pre-implantation mouse embryos. Mech Dev 125(3–4):270
Niwa H et al (2005) Interaction between Oct3/4 and Cdx2 determines trophectoderm differentiation. Cell 123(5):917

O'Sullivan CM et al (2004) Origin of the murine implantation serine proteinase subfamily. Mol Reprod Dev 69(2):126
Pearson H (2008) Making babies: the next 30 years. Nature 454(7202):260
Penner PS, Hull RT (2008) The beginning of individual human personhood. J Med Philos 32 (2):174
Perona RM, Wassarman PM (1986) Mouse blastocysts hatch in vitro by using a trypsin-like proteinase associated with cells of mural trophectoderm. Dev Biol 114(1):42
Peters PG Jr (2006) The ambiguous meaning of human conception. University of California, Davis Law Review 40:199
Piotrowska K et al (2001) Blastomeres arising from the first cleavage division have distinguishable fates in normal mouse development. Development 128:3739
Piotrowska-Nitsche K et al (2005) Four-cell stage mouse blastomeres have different developmental properties. Development 132:479
Plusa B et al (2005) Downregulation of Par3 and aPKC function directs cells towards the ICM in the preimplantation mouse embryo. J Cell Sci 118(3):505
Poueymirou WT et al (2007) F0 generation mice fully derived from gene-targeted embryonic stem cells allowing immediate phenotypic analyses. Nat Biotechnol 25(1):91
Primakoff P, Myles DG (2007) Cell–cell membrane fusion during mammalian fertilization. FEBS Lett 581(11):217
Rao M, Condic ML (2008) Alternative sources of pluripotent stem cells: scientific solutions to an ethical dilemma. Stem Cells Dev 17(1):1
Rubenfeld J (1991) On the legal status of the proposition that 'life begins at conception. Stanford Law Rev 4(3):599
Sowell ER et al (2003) Mapping cortical change across the human life span. Nat Neurosci 6(3):309
Stitzel ML, Seydoux G (2007) Regulation of the oocyte-to-zygote transition. Science 316:407
Takahashi K et al (2007) Induction of pluripotent stem cells from adult human fibroblasts by defined factors. Cell 131(5):861
Torres-Padilla M-E et al (2007) Histone arginine methylation regulates pluripotency in the early mouse embryo. Nature 445:214
Wang H et al (2008) Zonula occludens-1 (ZO-1) is involved in morula to blastocyst transformation in the mouse. Dev Biol 318(1):112
Worrad DM, Ram PT, Schultz RM (1994) Regulation of gene expression in the mouse oocyte and early preimplantation embryo: developmental changes in Sp1 and TATA box-binding protein, TBP. Development 120(8):2347
Yagi R et al (2007) Transcription factor TEAD4 specifies the trophectoderm lineage at the beginning of mammalian development. Development 134(21):3827
Yu J et al (2007) Induced pluripotent stem cell lines derived from human somatic cells. Science 318(5858):1917

Chapter 4
Complete Moles and Parthenotes Are Not Organisms

Nicanor Pier Giorgio Austriaco, O.P.

Abstract The complete hydatidiform mole is an embryo-like entity that is generated by an abnormal fertilization event. In the past, noted moralists have argued that hydatidiform moles are not embryos since these so-called embryos are substantially defective from the outset. More recently, however, other bioethicists have suggested that the human mole *may once have been* a human embryo. In response, I argue that the ontological status of moles, parthenotes, and other embryo-like entities that develop into tumors will depend upon two distinctions, the distinction between an active and a passive potential and the distinction between a whole and a part. I propose that an embryo-like entity that has an active potential to become a tumor in the whole (such entities would include complete hydatidiform moles and parthenotes) is a non-embryo, while an entity that only has an active potential to become a tumor only in the part (such entities would include partial hydatidiform moles) is an embryo, albeit a disabled one.

Keywords Complete hydatidiform mole • Embryos • Non-embryos • Organism • Parthenotes

4.1 Introduction

The complete hydatidiform mole is an embryo-like entity that is generated by an abnormal fertilization event. In human beings, moles develop into a disorganized mass that can become a germ-line tumor. In the past, noted moralists have argued

This chapter includes updated arguments initially made in my earlier essay, "Are teratomas embryos or non-embryos?: a criterion for oocyte-assisted reprogramming," Natl Cathol Bioeth Q 5 (2005): 697–706.

N.P.G. Austriaco, O.P.
Department of Biology, Providence College, Providence, RI, USA
e-mail: naustria@providence.edu

A. Suarez and J. Huarte (eds.), *Is this Cell a Human Being?*,
DOI 10.1007/978-3-642-20772-3_4, © Springer-Verlag Berlin Heidelberg 2011

that hydatidiform moles are not embryos since these so-called embryos are substantially defective from the outset. [Germain Grisez writes: “Many biologists and physicians going by appearance have believed these tumors to be embryos whose development had gone astray. But authoritative and recent examination of the question has led to the conclusion that these growths are simply tumors – rather disorganized but somewhat differentiating bundles of material deriving from an individual’s own body. Teratomas are not malformed embryos.” See Grisez (1970, p. 28). Benedict Ashley, OP, and Albert S. Moraczewski, OP, agree with Grisez: “Bedate and Cefalo argue from existence of hydatidiform moles and teratomas that genetic individuation is not sufficient since these entities arise from zygotes. This argument, however, seems definitively refuted since it is now known that the so-called ‘zygote’ in question is radically defective from the outset.” See Ashley and Moraczewski (1994).] More recently, however, other bioethicists have suggested that the human mole *may once have been* a human embryo. For instance, Pietro Ramellini has argued that an androgenetic complete mole (AnCHM) that arises from the fertilization of an enucleated egg with two sperm was once a human organism during a “premolar” stage (Ramellini 2006). So, are moles, parthenotes, and other teratoma-forming entities embryos or non-embryos?

In this paper, I argue that the ontological status of moles, parthenotes, and other embryo-like entities that develop into tumors will depend upon two distinctions: the distinction between an active and a passive potential and the distinction between a whole and a part. I propose that an embryo-like entity that has an active potential to become a tumor in the whole (such entities would include complete hydatidiform moles and parthenotes) is a non-embryo, while an entity that only has an active potential to become a tumor only in the part (such entities would include partial hydatidiform moles) is an embryo, albeit a disabled one. Finally, I suggest that an inner cell mass taken from an intact blastocyst is a non-organism that is not unlike an isolated cluster of pluripotent stem cells.

4.2 Hydatidiform Moles and Parthenotes: Biological Notes

There are two kinds of hydatidiform moles (for reviews, see Slim and Mehio 2007; Devriendt 2005). In a partial hydatidiform mole, two sperm fertilize a normal egg resulting in a conceptus with 69 chromosomes. The presence of three copies of each chromosome (triploidy) leads to death early in the pregnancy or, more rarely, in late-term loss of the abnormal fetus. Partial moles are characterized by disorganized placental and fetal growth. However, and this is important, the partial hydatidiform mole often presents itself as a tumor associated with a recognizable fetus.

In contrast, in a complete mole, either two sperm fertilize an enucleated egg (approximately 20% of complete moles) or a single sperm fertilizes an enucleated egg and then undergoes a duplication of its haploid genome (approximately 80% of complete moles). Thus, the product of conception has 46 chromosomes – the normal number – but all of them are derived from the father. The resulting mass that arises is made up solely of tissue derived from one type of embryonic cell type, the trophectoderm, though there is a report that *early* complete moles have cells derived from the inner cell mass (Zaragoza et al. 1997). In other words, the developed tumor is made up of exclusively placental tissue and there is no fetal tissue present. Complete hydatidiform moles always develop into teratomas.

Finally, a parthenote is an egg that has been activated to begin to divide and to develop in the absence of sperm [for a review, see Rougier and Werb (2001)]. There are many experimental procedures, mechanical, electrical, and chemical, which can artificially activate the mammalian egg in this way. In vitro, activated human oocytes have been able to develop to the 5-day, 100-cell blastocyst stage in a manner apparently indistinguishable at the gross morphological level from the early development of normal human embryos (Rogers et al. 2004). In vivo, there is evidence that human parthenotes develop into ovarian tumors [Oliveira et al. (2004); also, see Lee et al. (1997).] These observations may not be contradictory since a report that studied the development of mouse parthenotes in vitro suggests that in this species, activated eggs develop to what appears to be a blastocyst stage before becoming tumors (Hirao and Eppig 1997).

4.3 Hydatidiform Moles: Philosophical Analysis

So, are hydatidiform moles and parthenotes disabled embryos or non-embryos? Are they organisms or non-organisms? First, there is good reason to think that a partial hydatidiform mole is an abnormal human embryo not unlike those children born with a triploid genome. [Though there are reports of children born with a triploid genome, they do not survive beyond 10.5 months. For one case study, see Sherard et al. (1986).]

To explain, with partial hydatidiform moles, the tumor is often associated with a grossly abnormal but recognizable human fetus. The presence of this fetus suggests that with a partial mole, the extra copy of each gene distorts the development of the embryo. One manifestation of this abnormality is the development of a tumor. Or to put it another way, with a partial mole, an already abnormal embryo develops cancer.

On the other hand, there is also good reason to think that a complete hydatidiform mole is not an embryo. In brief, a complete mole does not have a complete *functional* genome. Thus, it lacks molecules ab initio that radically changes its developmental trajectory. Instead of becoming an organized structure of differentiated cells and tissues, a complete mole develops into a germ-line tumor of one cell type of placental origin.

To explain, recall that the complete mole inherits all of its 46 chromosomes from the father. This is significant because in mammalian organisms, a small number of genes (nearly 200 human genes in a recent count or 1% of the human genome [for details, see the paper Luedi et al. (2007); in the mouse, two significant papers have identified over 300 genes that are imprinted in the brain: Gregg et al. (2010a, b)] are inactive if they are inherited from a particular parent and not from the other. This is the phenomenon called genomic imprinting [for a review of genomic imprinting, see Kiefer (2007) and Wrzeska and Rejduch (2004)]. For example, the *IGF2* gene encoding insulin-like growth factor-2 is expressed if it is inherited from the father, but it is imprinted, i.e., rendered silent, if it is inherited from the mother [for details, see Chao and D'Amore (2008), Ohlsson (2004), also see Ohlsson et al. (1993)]. In contrast, the *IGF2R* gene encoding insulin-like growth factor-2 receptor is expressed if it is inherited from the mother, but it is imprinted if it is inherited from the father (Wutz et al. 1997). Thus, in effect, a complete mole lacks all the imprinted genes that are only active if they are inherited from the mother. Functionally, therefore, a complete mole has an incomplete genome. Significantly, there is evidence from mice that suggests that the imprinted genes that are nonfunctional in moles have an effect on the development of the whole embryo from the very beginning at the two-cell stage (Rappolee et al. 1992). The expression of imprinted genes has also been reported in human preimplantation embryos (Monk and Salpekar 2001; Salpekar et al. 2001). Thus, the complete mole is a living system that lacks molecules that are absolutely associated with a human embryo. [For a description of the systems perspective that is presupposed here, see Austriaco (2002, 2004).] Not surprisingly, because of the absence of these molecules that are only present when the genes that encode them are inherited from the mother, the complete hydatidiform mole cannot progress through the developmental stages associated with early human embryogenesis. Instead, it simply grows into a tumor often composed of only one or, at most, a few cell types of placental origin, though as we already noted above, there is a report that early complete moles have cells derived from the inner cell mass.

Together, the data suggest that the complete hydatidiform mole cannot be and is not any kind of unified organism ab initio. In other words, from the beginning, it is not an individual member of a particular biological species distinguished by a species-specific developmental trajectory that consists of the sequential and ordered appearance of differentiated cells and tissues. Therefore, it cannot be an organism. It is not an embryo.

Challenging this conclusion, Ramellini has suggested that as moles develop and become highly disorganized, they may pass through an initial stage of normality where they are normal or nearly normal human organisms that then become subject to disorganizing forces. More specifically, he suggests that a complete hydatidiform mole arises from a premolar entity that has organismic life. This premolar entity would be a normal human embryo that develops to the blastocyst stage. At this point, cells constituting the trophoblast of the blastocyst would initiate a molar transformation, killing the embryo.

To respond, we begin by noting that there is a distinction between an active and a passive potential. An active potential is actualized wholly from within. It is indicative of an entity's nature – its ontological status. For example, an acorn has an active potential to become an oak tree. In contrast, a passive potential is actualized from without. It requires the active causal intervention of an external agent in order to be realized. Thus, an acorn only has a passive potential to become a crucifix because it would need the agency of a master craftsman in order to realize this end.

Given this distinction, the transformation of a premolar entity into a complete hydatidiform mole that becomes a tumor could potentially be described in three ways. First, we could say that the complete mole is a human organism that has an active potential to become a tumor. This would be incoherent. By nature, human embryos do not become tumors. Thus, they cannot have an active potential to become teratomas. Second, we could say that a complete mole could have been a human embryo that had a passive potential to become a tumor. By definition, however, a passive potential needs an external agent in order to be realized, and it is clear from his narrative that Ramellini is referring not to external but to internal disorganizing forces arising from non-expressed or inappropriately expressed genes. He writes: "The exact underlying mechanism [of the molar transformation] is still largely unknown, but both paternally and maternally imprinted genes are involved, with overexpression of certain paternal, and lack of expression of certain maternal genes." [Ramellini (2006); however, we should also acknowledge that there is evidence that normal mammalian embryos have a *passive* potential for teratoma formation. For example, if a normal mammalian embryo is transplanted under the kidney capsule of an athymic mouse, it can develop into a teratoma. Here, it is likely that the abnormal physiological environment of the kidney capsule kills the embryo by transforming it into a tumor. For technical details, see Anderson et al. (1996)]. Thus, and third, we have to say for Ramellini a complete mole is a human embryo where one part of the whole had an "active" potential to become a tumor. (Properly speaking, since active potentials are manifestations of substantial forms, which are principles of organization, there can be no active potential for tumor formation, a process that involves disorganization rather than organization. Therefore, it is more proper to say that a tumor arises in a human being because of defect in the *material cause* that affects a part of the human embryo. To put it another way, a tumor arises because the formal cause, the human soul, is unable to properly realize the potencies in part of the material cause. This defect leads to the abnormal actualization of hidden potencies in the material cause that manifest themselves as the disordered growth we call a tumor. For this clarification, I am indebted to Professor Michel Bastit of the University de Bourgogne in Dijon, France.) In the course of development, this part develops into the tumor that radically distorts the overall trajectory of the embryo of which it is a part, killing it.

Given this analysis, we can respond by asking the following question: In a complete hydatidiform mole, do the genetic defects that distort its developmental trajectory leading to tumor formation affect the whole mole or only a part of it? If the imprinting or silencing defects affect the whole, then tumor formation is the actualization of an "active" potential that reflects the nature of the mole. (Again, properly speaking, there can be no active potential for tumor formation, a process that involves disorganization rather than organization. In this case, it is more correct to say that a tumor arises because of a defect in the *material cause* that affects the entire cell mass. To put it another way, a tumor arises because the formal cause, the human soul, is unable to properly inform the defective matter. This failure of ensoulment leads to the abnormal actualization of hidden potencies in the material cause that manifest themselves as the disordered growth we call a tumor.) On the other hand, if the imprinting defects affect a part of the mole, then tumor formation is simply the distortion of a once normal embryo. Though to the best of my knowledge no scientific experiments on human embryos have been performed to directly address this question – incidentally, experiments that would be morally reprehensible – there is scientific evidence from developing mice that suggests that the absence of parentally imprinted genes impacts the development of the whole embryo and not only a part, even at the two-cell stage [Rappolee et al. (1992), also see the paper by Walsh et al. (1994)]. This suggests that tumor formation is a manifestation of the nature of the complete mole. Therefore, a complete mole is not an organism because an organism does not become a tumor. In contrast, as already noted above, the evidence from partial moles suggests that tumor formation is a manifestation of a defect in a part where one part of the embryo becomes a tumor in the context of the abnormal development of the fetus. Thus, a partial mole is an embryo, albeit a disabled one.

Finally, a comment: Based on the published report that identified cells derived from the inner cell mass in early complete hydatidiform moles, some bioethicists assumed that a complete hydatidiform mole undergoes normal human development to the blastocyst stage. At this point, the trophoblast would overwhelm and destroy the adjacent inner cell mass. There is no direct empirical evidence for this. In fact, a recent report describing the case of a complete hydatidiform mole, a placenta, and a coexisting fetus derived from a single in vitro fertilized oocyte challenges this assumption (Hsu et al. 2008). In this case, the complete mole and the intact fetus developed from a single 12-celled embryo. Though the exact mechanism behind this case of a complete mole/placenta/fetus combination is unclear, what is clear is that the cells that gave rise to the complete mole coexisted with the cells that gave rise to the intact embryo at the beginning of the pregnancy and that they developed into a tumor without overwhelming and killing the embryo proper. This would not be expected if the assumption that a molar trophoblast would have killed an adjacent embryo had been true. Therefore, I favor an alternative explanation for the presence of ICM-derived tissue in early complete moles: A complete hydatidiform mole is able to generate tissues of different cell types in a disorganized fashion early in its development.

4.4 Parthenotes: Philosophical Analysis

At this point, it is important to note the parallels between complete hydatidiform moles and parthenotes. Recall that a complete mole is an embryo-like entity that results from an abnormal fertilization event where two sperm fertilize an enucleated egg. Thus, a complete mole has the normal number of 46 human chromosomes, but they are all derived from the father. In contrast, a human parthenote results from an abnormal "fertilization" event where an egg has been activated to begin dividing.

In the process of activation, the haploid egg duplicates its 23 chromosomes. Thus, the parthenote too has the normal number of 46 human chromosomes, but here, they are all derived from the mother. Not surprisingly, therefore, mammalian parthenotes, such as complete moles, have a defective genome. In this case, they lack all the imprinted genes that are only active if they are inherited from the father. Not surprisingly, therefore, our analysis of the ontological status of parthenotes parallels our earlier analysis of the ontological status of complete hydatidiform moles.

For our analysis, the definitive question that arises is the following: Does the absence of the maternally imprinted gene products that are only produced when genes are inherited from the father impact the system dynamics of the whole parthenote ab initio substantially changing it so that the parthenote becomes a tumor, or does their absence only lead to a defective part of an embryo that becomes a tumor eventually killing the whole?

In response, we return to the key study, already mentioned above, that has shown that the absence of both paternally and maternally imprinted genes impacts the development of the embryo from the very beginning at the two-cell stage (Rappolee et al (1992). Moreover, the absence of the maternally imprinted molecules impacts that parthenote at the level of the whole since the scientific evidence suggests that the imprinted genes regulate the overall number of cells that develop in the blastocyst. In addition, another study has demonstrated that the organization of a parthenote differs from the organization of a normal embryo from the very start: In normal development, when the single-celled mouse zygote divides into two cells, these two cells, called blastomeres, are already not identical. One of the two cells divides ahead of its sister and tends to contribute most of its cellular descendents to the ICM that generates the embryo proper, which will develop into the baby's body, whereas the other, later dividing cell, contributes cells predominantly to the extra-embryonic tissue including the placenta, which will develop into the afterbirth [for details, see Piotrowska et al. (2001).] However, in contrast, when a single-celled mouse parthenote divides into two cells, these two cells do not behave in the way that normal blastomeres would behave [for details, see Piotrowska et al. (2001).] The first cell that divides does not necessarily contribute its descendents to the ICM. This is a small but significant difference in organization that points to the difference between the parthenote and the normal embryo at the very earliest stages of the development. In sum, all of these data suggest that the forces that lead to tumor formation are already present at the earliest stages of embryonic development. In other words, the system dynamics of the parthenote as a whole already differs from the system dynamics of the normal embryo since normal embryos do not become teratomas. Like the complete mole, a parthenote is not an embryo.

But what about blastocyst formation? As we noted above, there is evidence that human parthenotes have been able to develop to the blastocyst stage in vitro. If the parthenote is not an embryo, why does it develop at least until the blastocyst stage? There is a ready explanation for the development of blastocyst structures in the parthenote. Numerous studies have shown that the egg contains molecules from the mother that can compensate for defects in the embryo's genome. For instance, embryos lacking the gene for E-cadherin, an essential molecule that glues cells together, are able to maintain their integrity until the blastocyst stage because the mother provides it with her E-cadherin. However, when the store of maternally derived E-cadherin is depleted, the embryo's cells dissociate and the embryo collapses (Larue et al. 1994). This defect is evident at the molecular level from the very beginning as levels of maternally derived E-cadherin molecules gradually decrease, but the morphological effects take time to manifest themselves. In the same way, it is not surprising that the absence of the maternally imprinted molecules in the parthenote does not completely manifest itself until the blastocyst stage. The molecular defect within the parthenote is temporarily masked by the molecules inherited from the mother. This, however, does not detract from the reality that the parthenote, in itself, is a teratoma-forming entity, a non-embryo, from the very beginning that has an active potency to become a tumor.

4.5 Conclusion

Finally, I would like to close with a brief word about the isolated inner cell mass. The earliest sign of cell differentiation during human development occurs during the transformation of the morula to the blastocyst with the appearance of the trophoblast and inner cell mass (ICM). The trophoblast gives rise to the embryonic contribution to the placenta, while the inner cell mass generates the embryo's body proper and its extraembryonic tissues [Gardner and Beddington (1988); also see the molecular analysis by Adjaye et al. (2005).] Is the ICM an organism or a non-organism? It is clear that an isolated ICM cannot continue developing into a mature organism. When placed in culture, these ICM cells eventually die, though some of them will be transformed into pluripotent ES cells via a still uncharacterized stochastic mechanism [for a representative paper, see Lerou et al. 2008.] Thus, I propose that the ICM is a non-organism that is not unlike a clump of pluripotent stem cells. ICM cells retain the ability to generate all the cell types and tissues of the mature organism, but they are unable to do so in an ordered and sequential way, the way that characterizes a *bona fide* organism, the totipotent embryo.

References

Adjaye J, Huntriss J, Herwig R, BenKahla A, Brink TC, Wierling C, Hultschig C, Groth D, Yaspo ML, Picton HM, Gosden RG, Lehrach H (2005) Primary differentiation in the human blastocyst: comparative molecular protraits of inner cell mass and trophectoderm cells. Stem Cells 23:1514–1525

Anderson GB, BonDurant RH, Goff L, Groff J, Moyer AL (1996) Development of bovine and porcine embryonic teratomas in athymic mice. Anim Reprod Sci 45:231–240

Ashley B, Moraczewski AS (1994) Is the biological subject of human rights present from conception? In: Cataldo PJ, Moraczewski AS (eds) The fetal tissue issue. The Pope John XXIII Center, Braintree, MA, pp 33–59
Austriaco N (2002) On static eggs and dynamic embryos: a systems perspective. Natl Cathol Bioeth Q 2:659–683
Austriaco N (2004) Immediate hominization from the systems perspective. Natl Cathol Bioeth Q 4:719–738
Austriaco N (2005) Are teratomas embryos or non-embryos?: a criterion for oocyte-assisted reprogramming. Natl Cathol Bioeth Q 5:697–706
Chao C, D'Amore PA (2008) IGF2: epigenetic regulation and role in development and disease. Cytokine Growth Factor Rev 19:111–120
Devriendt K (2005) Hydatidiform mole and triploidy: the role of genomic imprinting in placental development. Hum Reprod 11:137–142
Gardner RL, Beddington RS (1988) Multi-lineage stem cells in the mammalian embryo. J Cell Sci Suppl 10:11–27
Gregg C, Zhang J, Butler JE, Haig D, Dulac C (2010a) Sex-specific parent-of-origin allelic expression in the mouse brain. Science 329:682–685
Gregg C, Zhang J, Weissbourd B, Luo S, Schroth GP, Haig D, Dulac C (2010b) High-resolution analysis of parent-of-origin allelic expression in the mouse brain. Science 329:643–648
Grisez G (1970) Abortion: the myths, the realities, and the arguments. Corpus Books, New York
Hirao Y, Eppig JJ (1997) Parthenogenetic development of Mos-deficient mouse oocytes. Mol Reprod Dev 48:391–396
Hsu CC, Lee IW, Su MT, Lin YC, Hsieh C, Chen PY, Tsai HW, Kuo PL (2008) Triple genetic identities for the complete hydatidiform mole, placenta and co-existing fetus after transfer of a single *in vitro* fertilized oocyte: Case report and possible mechanisms. Hum Reprod 23:2686–2691
Kiefer JC (2007) Epigenetics in development. Dev Dyn 236:1144–1156
Larue L, Ohsugi M, Hirchenhain J, Kemler R (1994) E-cadherin null mutant embryos fail to form a trophectoderm epithelium. Proc Natl Acad Sci USA 19:8263–8267
Lee GH, Bugni JM, Obata M, Nishimori H, Ogawa K, Drinkwater NR (1997) Genetic dissection of susceptibility to murine ovarian teratomas that originate from parthenogenetic oocytes. Cancer Res 57:590–593
Lerou PH, Yabuuchi A, Huo H, Takeuchi A, Shea J, Cimini T, Ince TA, Ginsburg E, Racowsky C, Daley GQ (2008) Human embryonic stem cell derivation from poor-quality embryos. Nat Biotechnol 26:212–214
Luedi PP, Dietrich FS, Weidman JR, Bosko JM, Jirtle RL, Hartemink AJ (2007) Computational and experimental identification of novel human imprinted genes. Genome Res 17:1723–1730
Monk M, Salpekar A (2001) Expression of imprinted genes in human preimplantation development. Mol Cell Endocrinol 183:S35–S40
Ohlsson R (2004) Loss of IGF2 imprinting: mechanisms and consequences. Novartis Found Symp 262:108–21
Ohlsson R, Nyström A, Pfeifer-Ohlsson S, Töhönen V, Hedborg F, Schofield P, Flam F, Ekström TJ (1993) IGF2 is parentally imprinted during human embryogenesis and the Beckwith-Wiedemann syndrome. Nat Genet 4:94–97
Oliveira FG, Dozortsev D, Diamond MP, Fracasso A, Abdelmassih S, Abdelmassih V, Gonçalves SP, Abdelmassih R, Nagy ZP (2004) Evidence of parthenogenetic origin of ovarian teratoma: case report. Hum Reprod 19:1867–70
Piotrowska K, Wianny F, Pedersen RA, Zernicka-Goetz M (2001) Blastomeres arising from the first cleavage division have distinguishable fates in normal mouse development. Development 128:739–748
Ramellini P (2006) Life and organisms. Libreria Editrice Vaticana, Vatican City
Rappolee DA, Sturm KS, Behrendtsen O, Schultz GA, Pedersen RA, Werb Z (1992) Insulin-like growth factor II acts through an endogenous growth pathway regulated by imprinting in early mouse embryos. Genes Dev 6:939–952

Rogers NT, Hobson E, Pickering S, Lai FA, Braude P, Swann K (2004) Phospholipase C-zeta causes Ca^{2+} oscillations and parthenogenetic activation of human oocytes. Reproduction 128:697–702
Rougier N, Werb Z (2001) Minireview: parthenogenesis in mammals. Mol Reprod Dev 59: 468–474
Salpekar A, Huntriss J, Bolton V, Monk M (2001) The use of amplified cDNA to investigate the expression of seven imprinted genes in human oocytes and preimplantation embryos. Mol Hum Reprod 7:839–844
Sherard J, Bean C, Bove B, DelDuca V Jr, Esterly KL, Karcsh HJ, Munshi G, Reamer JF, Suazo G, Wilmoth D, Dahlke MB, Weiss C, Borgaonkar DS (1986) Long survival in a 69, XXY triploid male. Am J Med Genet 25:307–312
Slim R, Mehio A (2007) The genetics of hydatidiform moles: new lights on an ancient disease. Clin Genet 71:25–34
Walsh C, Glaser A, Fundele R, Ferguson-Smith A, Barton S, Surani MA, Ohlsson R (1994) The non-viability of uniparental mouse conceptuses correlates with the loss of the products of imprinted genes. Mech Dev 46:55–62
Wrzeska M, Rejduch R (2004) Genomic imprinting in mammals. J Appl Genet 45:427–33
Wutz A, Smrzka OW, Schweifer N, Schellander K, Wagner EF, Barlow DP (1997) Imprinted expression of the Igf2r gene depends on an intronic CpG island. Nature 389:745–749
Zaragoza MV, Keep D, Genest DR, Hassold T, Redline RW (1997) Early complete hydatidiform moles contain inner cell mass derivatives. Am J Med Genet 70:273–277

Chapter 5
Embryos Grown in Culture Deserve the Same Moral Status as Embryos After Implantation

Joachim Huarte and Antoine Suarez

Abstract Reprogramming somatic adult cells makes it possible to obtain induced pluripotent stem cells (iPSCs) without the necessity of destroying human embryos. Nonetheless, this technique gives rise to unexpected ethical challenges. In particular, the possibility of using reprogramming to create human gametes and eventually embryos has provoked arguments negating the moral status of human embryos that can be grown in culture.

To face these new ethical challenges we introduce the concept of *proper biological potential* for developing a body exhibiting human architecture and spontaneous motility and further propose that personhood can be ascertained on the basis of this *proper biological potential*. We then argue that the embryo's *proper biological potential* for developing the neural activity responsible for fetal motility is not determined by implantation and so embryos grown in culture deserve the same moral status as the embryos after implantation. Additionally, we articulate our philosophical reasons for our support of the clinical criteria for defining death and our support of the criterion of DIANA insufficiencies (insufficiencies that Directly Inhibit the Appearance of Neural Activity) for distinguishing between disabled embryos and nonembryos.

J. Huarte
Department of Genetic Medicine and Development, University of Geneva Medical School, 1, rue Michel-Servet, 1211 Geneva 4, Switzerland (At the time of manuscript summission)

Social Trends Institute/Bioethics, Barcelona, Spain

Swiss Society of Bioethics, Zurich, Switzerland
e-mail: joachim.huarte@gmail.com

A. Suarez (✉)
The Institute for Interdisciplinary Studies, Berninastr. 85, 8057 Zurich, Switzerland

Social Trends Institute/Bioethics, Barcelona, Spain

Swiss Society of Bioethics, Zurich, Switzerland
e-mail: suarez@leman.ch

A. Suarez and J. Huarte (eds.), *Is this Cell a Human Being?*,
DOI 10.1007/978-3-642-20772-3_5, © Springer-Verlag Berlin Heidelberg 2011

Keywords DIANA insufficiencies • Neural activity • Proper biological potential • Spontaneous movements

5.1 Introduction

The perception that complex cellular interactions within tissues and organs are altered or impaired in human diseases has caused much of modern medicine to become more interested in therapies utilizing cells as reparative tools. Stem cells have been the cells of choice because of their relative indeterminacy and special capacity to divide and eventually give rise to specialized cells capable of forming functional tissue structures and organs. In the late 90s, great enthusiasm arose from what since has become a contentious topic: the derivation of stem cells from human embryos. Embryonic stem cells (ESCs) carried great promise because such cells had the potentiality of giving rise to any kind of cell in the body. However, the production and use of human ESCs immediately proved to be controversial, as the methods then available required the destruction of human embryos in order to obtain ESCs. As a result, many researchers sought alternative ways for ESC production.

In 2006, Shinya Yamanaka and coworkers proposed a technique for deriving pluripotent stem cells by prompting adult cells, through a relatively simple method using four genetic factors, to revert to an undifferentiated state (Check 2006; Takabashi and Yamanaka 2006). The resulting cells, dubbed "induced pluripotent stem cells (iPSCs)," possess the same kind of potentiality as ESCs, with the added advantage that their use averts the destruction or use of human embryos. Advances in this area of research have been quite rapid and presently we are able to induce pluripotency in a variety of human cells (For recent reviews see Plath and Lowry 2011; González et al. 2011). Though this method clearly avoids the caveat of destroying human embryos, it raises other ethical concerns.

As the contributions in *Nature*, July 17, 2008, make clear (Nature 2008), the method could be used to reprogram human somatic cells so that they become human germ cells and give rise ultimately to human sperm and ova. Thus, iPSCs could be used to create human gametes and eventually a human embryo. Given that the production of human beings in vitro has been, until now, limited by the ability of scientific researchers, fertility doctors, and potential parents to obtain human gametes at efficient cost, the use of Yamanaka's method could have deeply ethical implications, especially for those who find the production of human beings in a laboratory setting to be an affront to human dignity.

Davor Solter takes things straight to the point in the aforementioned issue of *Nature*: "Today you can't experiment on human embryos because it's considered morally repugnant – and they are difficult to get. If embryos could be grown in culture like any other cell line, this latter problem would disappear. [. . .] They'd become like any other type of cell line. They would become objects and would be used as objects" (Nature 2008, p. 260). Solter's claim shows that despite recent advances in stem cell research that could both allow important research to go forward and satisfy the moral convictions of many people we are still facing a hardened denial of any kind of moral status to the human embryo.

Remarkably, a number of authors like Solter do not question that a child and the adult into which the child can develop are the same human person and deserve the same moral status. How then are they led to deny that embryos in culture are the same person and deserve the same moral status as the child and the adult into which these embryos can develop if they were to be implanted?

In this respect, it is important to see how Nobel laureate Christiane Nüslein-Volhard argues: "From a biological point of view, nearly nothing is more discontinuous than the process [of implantation], by which an embryo comes into direct cell contact with another individual. In the fertilized egg there is the complete genetic program. But for the program's processing, the embryo requires the intensive interaction, the symbiosis with a second organism – that of the mother [. . .]. Thus the developmental potential is first there with the implantation and becomes completed at birth" (Nüsslein-Volhard 2004).

Much in the same line of thinking Hans Schöler states: "It is true that a human being can originate from such a blastocyst. However this embryo, in such an early stage of development, is a little cell-ball, tiny like a needle tip. It does not yet show any resemblance to a human being. To its further grow and development, it needs the implantation into the uterus of a woman. If this does not succeed, it perishes within some days" (Schöler 2007).

One can certainly respond that these arguments lack coherence because development, in particular of the brain, is not completed at birth but continues after it and, therefore, one could as well argue that an infant is clearly not the same as an adult. Scholars endorsing the embryo's rights often give this sort of response and we think it adequate. Nonetheless, the quotations earlier show that apparently experienced researchers earnestly back the following assumption: the embryo lacking interaction with the maternal organism (or an equivalent growth culture) cannot reach the full developmental potentiality and, therefore, the moral status of personhood. We think it is worth discussing whether this assumption can be considered correct. Such a discussion may help promote a more open-minded attitude among the scientific community towards the possibility of embryonic rights.

Another major ethical concern relates to the generation of fully iPSCs-derived mice through tetraploid complementation. In this context one asks whether reprogramming techniques may become a new and more efficient sort of cloning than the conventional one (Denker 2009). If this were the case, the usual ethical arguments against cloning would also apply to the combination of iPSCs with tetraploid embryos. However, there is another aspect. Fully iPSCs-derived mice demonstrate that it is possible in principle to revert somatic cells to a colony of cells from which a healthy individual could be born, that is, inducing totipotency and not only pluripotency. Thus, we think it is also worth discussing how to determine whether a particular reprogramming technique produces cells sharing embryo status.

This chapter is organized as follows:

In Sect. 5.2, we introduce the concept of "*proper biological potential* to unfold the neural activity responsible for spontaneous movements of a human body," and argue in favor of the philosophical principle that the presence of this *proper biological potential* in an organism consisting of human cells allows us to state that this organism is animated by a spiritual soul and is therefore a human person.

In Sect. 5.3, we compare our philosophical position to that of Aristotle and Thomas Aquinas. On the one hand we argue, in agreement with these authors, that it is possible to have an organism made of flesh (cells) of human origin that is not animated by a spiritual soul. On the other hand we argue, in disagreement with them, that each organism made of flesh of human origin and sharing specific animal life (capability for spontaneous motility and sense) is animated by a spiritual soul.

In Sect. 5.4, we show that our philosophical principle is consistent with the standard clinical criteria for defining death and therefore that these criteria can be considered sufficient for establishing that a whole brain-dead organism is not a person. By contrast, the interpretation of "brain death" as a loss of the ability of the body's parts (organs and cells) to function together as an integrated whole is actually incomplete, since one can hardly deny that a brain-dead organism possesses a certain degree of integrated functioning. The concept of "organism" without further specification does not provide a good enough criterion to ascertain animation by a spiritual soul.

In Sect. 5.5, we summarize some scientific essentials about embryonic development, and then argue, in Sect. 5.6, that the embryo's *proper biological potential* of developing neural activity is not determined by the interaction with the maternal organism. On the basis of these scientific observations, we conclude that a human embryo in culture is animated by a human soul independently of whether the embryo becomes implanted or not.

In Sect. 5.7, we stress that distinguishing between embryos and nonembryos remains an urgent task and to this aim propose criteria for such a distinction based on *DIANA insufficiencies* (insufficiencies that Directly Inhibit the Appearance of Neural Activity) (Suarez et al. 2007). DIANA insufficiencies are defects or alterations that frustrate the development of the cellular path leading directly from the zygote to the differentiation of brain cells. We argue that human organisms exhibiting an embryo-like developmental pattern but bearing a DIANA insufficiency can morally be compared to a human organism fulfilling the clinical criteria of brain death. This comparison does not hold for other anomalies frustrating the development on other secondary developmental branches, as for instance alterations generating defective extraembryonic tissues or cardiac muscle. In this context, we stress again that the concept of "organism" bears ambiguities as a criterion to decide on the moral status of cell entities.

As it will appear in Sect. 5.8, our proposed criteria of *proper biological potential* and DIANA insufficiencies depend on the assumption that the inner cell mass (ICM) is not equivalent to the colony of ESCs deriving from it. Thus, in order to settle certain arguments it may be of interest to test this assumption by performing a complementation experiment in animal models, which to our knowledge has not yet been done.

Section 5.9 will highlight that this experiment is important to prohibit the reprogramming of adult cells that would produce ICM-equivalent colonies of iPS cells since such techniques would actually produce embryos and be a new form of cloning.

5.2 *Proper Biological Potential* to Become a Body with Both Human Architecture and Spontaneous Motility as an Observable Basis for Ascertaining Animation by a Spiritual Soul

Determining the moral status of an organism requires some set of observable criteria. In other words, when ascertaining moral status, we move first from what is observable (sets of empirical or sense data) and on the basis of that make a judgment as to the presence or absence of a spiritual soul in a concrete body. Based on this principle, we argued in previous articles (Huarte and Suarez 2004; Suarez 1993) that any organism possessing the biological potential to develop a body with a human architecture and spontaneous motility is a human person. We restate and improve this argument in the present section.

Anyone who believes in a spiritual soul necessarily assumes that his or her soul is somehow involved in governing certain movements of his or her body. When I am speaking to someone or writing a letter, I have the immediate internal experience that I am controlling the movements of my lips, hands, and eyes through the spiritual powers of my soul, i.e., my free will and intellect.

But how do I conclude that the body of someone other than me is animated by a spiritual soul as well? When I gaze my eyes upon a colleague speaking to me during lunch, I conclude that the human being sitting in front of me is animated by a spiritual soul because he (a) has the same specific form (or shape) as my body, and (b) this form exhibits movements like the movements I make for expressing my thoughts, emotions, and claims for rights. Mirror neurons are the neurophysiologic correlate of this philosophical supposition, which is to say that the same class of neurons that fire in my brain when I move my lips also fire when I observe someone else moving his lips (Rizzolatti and Craighero 2004).

Movements of the human body like those often involved in expressing one's thoughts and wishes to other persons we call *spontaneous* movements. They may be arm, leg, or lip movements; head flexion and rotation; eye or breathing movements. Even if they are often unconscious and unintentional, they are potentially will-directed movements, that is, they are movements that can always be directed by the will when chosen. Spontaneous movements can be considered voluntary acts, and are actually ruled by the will even if they do not always originate in a conscious way and are not always immediately guided by some intention. I regulate the movement of my lips through my will when I move them consciously and when I move them unconsciously. It is a merit of Benjamin Libet's experiments to have shown that voluntary acts are not necessarily conscious ones (Libet 1985, 1992, 2002).

In contrast to these movements of the lips, tongue, eyes, fingers, etc., it is not possible to use one's "heart beat" to communicate with another person. Though this kind of movement is not caused by some stimulus external to the body, it is highly programmed according to one's biological makeup and thus cannot be brought under the immediate control of the will. These types of movements, like heart beating, we call *autonomous* movements. Another kind of movement that escapes the control of the will is a nerve *reflex* reaction that has an external cause. Neither autonomous movements nor reflexive ones are meant to fall under what we have termed "spontaneous" movements.

According to Aristotle "acts done by reason of anger or appetite are not rightly called involuntary," and it is appropriate to consider that nonhuman animals and children "act voluntarily" (Aristotle NE, MA). Continuing this line of thinking we assume that the concept of "voluntary" can be extended to our *spontaneous* movements as well, although most often they happen in an unconscious way or without deliberation. Our *autonomous* movements correspond to what Aristotle called "involuntary movements" like "the motions of the heart and the privy member" (Aristotle MA).

Steering the spontaneous movements of the human body is something intrinsic to the operation of expressing one's thoughts and wishes to other persons. We consider it appropriate, therefore, to assume that these movements directly reveal the spiritual powers of the human soul and accept that the neural activity responsible for their regulation is intrinsically governed by the soul. Nonetheless, we don't reduce the soul to the ensemble of the material deterministic properties of the brain (Suarez 2011).

By contrast, we do not associate the activity of the human soul directly with the *autonomous* activity of the heart, liver, or kidneys, even when a lethal injury to these organs ultimately causes the loss of spontaneous movements. Consequently, transplantation of the heart, for example, cannot be considered equal to "transplanting" the human soul from one body to another. Neither do we associate the activity of the soul directly to nerve *reflex* reactions.

Consider a human body B that receives a foreign heart and call B' the body with the new heart. It is today generally assumed that B and B' share the same personal identity. The fact that B' and B have different hearts does not prevent conservation of personal identity. In the same line of thinking, we establish *conservation of the personal identity* during bodily growth as follows: consider a cell entity or a cell layer X that develops into a body B exhibiting human architecture and spontaneous motility. Suppose that this development requires the interaction of X with other cell clusters or cell layers in its environment. Suppose, however, that the neural centers of B responsible for its spontaneous activity derive exclusively from cells contained in X; suppose also that equivalent neural centers cannot derive from the cell entities or cell layers interacting with X. Then we state that X has the *proper biological potential* to develop into B and shares the same personal identity (is animated by the same spiritual soul) as B. Other cell clusters or cell layers interacting with X share the status of the heart or skin of B, that is, they could be replaced without loss of personal identity. Thus, we reach the principle: any cell entity or cell layer possessing the *proper biological potential* to develop a body exhibiting human architecture and spontaneous motility is animated by a spiritual soul. It is in this sense that our concept of "minimal developmental potential" in previous articles should be understood (Huarte and Suarez 2004; Suarez 1993).

Benedict Ashley and Albert Moraczewski argue in much the same way, although not relying on spontaneous movements, but on *sensual perception*:

> Yet human intelligence is the weakest of all possible kinds of intelligence and hence is always dependent on sense cognition for all it can know. This sense data must be acquired by the external sense organs and then processed by some internal sense organ. When Aristotle and Aquinas wrote, medical science held that this internal sense organ was the heart because it was the most energetic part of the body, but now we know that this organ is in fact (at least in adults) the human brain... The principal organ required by his argument need not be one that here and now is capable of acting as the instrument of intellection. We are human persons even when we are asleep or comatose from a brain injury... Thus the primary organ that is required in the fetus for its intellectual ensoulment is not the brain as such but a primary organ capable of producing a brain with the capacity for intellectual cognition in the body at some appropriate phase of the human life cycle. (Ashley and Moraczewski 2001)

We think that the two criteria used for ascertaining the presence of a spiritual soul – sensory perception and spontaneous motility – cannot actually be separated from each other. Nevertheless, that analysis is beyond the scope of this chapter. Notice that, in any case, the practical standard we have for indicating that a fetus has brain function (and even sensory perception) is observing that the fetus exhibits spontaneous motility.

5.3 The *Proper Biological Potential* of Spontaneous Motility and the Hypothesis of Delayed Hominization

We would like to stress that the philosophical principle formulated in the preceding section does not mean that the human soul first "enters the body" when fetal motility or neural activity appears. The meaning of our premise is the following: any human body which shows spontaneous movements was spiritually ensouled from the very beginning of the process of cell division. In other words, a human cell that has the *proper biological potential* for reaching the stage of spontaneous movements is a spiritually animated human embryo and a person. Therefore, our premise does not imply at all that the human person undergoes delayed hominization during embryogenesis. However, it is interesting to note that the hypothesis of delayed hominization, as for instance formulated by Thomas Aquinas interpreting Aristotle (Jones 2004; Rodríguez-Luño 2006), is not only a consequence of "archaic biology," but is based on two philosophical assumptions worthy of being explicitly considered.

According to the theory of delayed hominization, in the generation of a man the embryo first lives the life of a plant through the vegetal soul; next, this form is removed by corruption and the embryo acquires, by a sort of new generation, a sensitive soul and lives the life of an animal; finally, this animal soul is in turn removed by corruption, and the ultimate and complete form is introduced. This is the rational soul, which comprises within itself whatever perfection was found in the previous forms (Aquinas CTh).

This view implies that there can be organisms or systems made of flesh or biological matter from human origin (human cells, in terms of today's biology) that are not animated by a spiritual soul. Among such organisms some are supposed to share features of vegetal life only (animation by a vegetative soul), and others specific features of animal life as well (animation by a sensitive soul). In addition to these philosophical assumptions, Aristotle and Aquinas advocate the biologically "archaic" view that it is cardiac, not neural activity, which is responsible for the specific characteristics of animal life (see the quotation of Ashley and Moraczewski in Sect. 5.2).

In agreement with Aristotle and Thomas Aquinas, we endorse the possibility of an organism made of human cells and sharing integrative functions like those of plants (growth, circulation, bearing offspring) that is not spiritually ensouled and therefore is not a person. In disagreement with them, however, we reject that an organism or biological system made of human cells and sharing animal life could be animated by a nonspiritual soul. That is, we accept that the spiritual soul is the unique form of the human body, but reject that a body made of human cells and sharing animal life can have a form that is not a spiritual one. And also in disagreement with Aristotle's and Aquinas' biology, we assume that the organ responsible for specific animal life is the brain and not the heart.

We think that one cannot appropriately decide when the human person begins and dies on the basis of principles like "movement," "life," "organization," "consciousness," or "thinking." Instead, one needs to derive the beginnings of human life based on their observation of a *body* that exhibits human architecture and spontaneous movements (e.g., those bodily movements I use to communicate with others, in particular to claim for my rights). It is only because we realize that a determined body is a *living human animal* (and not only a human "organism") that we conclude that it is animated by a spiritual soul and is therefore a human person. And we further conclude that the *proper biological potential* for bringing about such a body provides the criterion for ascertaining whether a cell entity is animated by a spiritual soul and has the moral status of a person.

In the following sections, we show how this criterion applies in different cases.

5.4 Implications for the Definition of Death

In *clinical praxis*, one declares a body brain dead primarily on the basis of clinical tests (so-called brain death criteria) that demonstrate that the body has "irreversibly" lost the capacity to perform determined spontaneous movements, for example, breathing, eye, and leg movements. The loss of the capacity for spontaneous motility goes along with the disappearance of the neural activity (including brainstem function) responsible for such movements (Laureys 2005). We would like to insist that in the clinical praxis actually one determines death on the basis of concrete operational clinical criteria allowing us to determine the irreversible loss of the capacity for certain movements, and not on the basis of a general definition requiring a determination that "all functions of the entire brain, including the brain stem" have irreversibly ceased.

In this sense, by recognizing that a brain-dead organism is not a human person one clearly associates personhood (or in philosophical terms, the activity of a human soul) with the *proper biological potential* for unfolding spontaneous motility. In particular, one correlates the activity of a human soul with neural activity as such activity reveals the capacity for spontaneous movements and, in particular, spontaneous breathing. This means that the standard clinical criteria for defining death are consistent with our premise for ascertaining personhood. For on the one hand, one must consider that the patient in a persistent vegetative state is ensouled by a spiritual soul, since he or she still exhibits extended spontaneous motility, in particular, spontaneous breathing (Laureys 2005). On the other hand, a brain-dead organism that exhibits a beating heart and blood circulation but has *irreversible* cessation of the neural activity responsible for spontaneous motility no longer has the *proper biological potential* for unfolding such activity. According to the philosophical premise we have adopted such an organism is not a person.

We would like to stress that our view is consistent with defining the death of a human person as a "loss of the integrative capacity of the organism." However, we suggest that there are some grounds for fine-tuning that assertion to mean the integrative capacity that makes an organism "capable of animal behavior." This is, in fact, what the standard clinical criteria of death are actually attempting to ascertain. In other words, the implicit philosophical premise from which these criteria are derived appears to be consistent with our philosophical categories.

On the contrary, using only "organization" or "integration" without further specification as a criterion for ascertaining the presence of a human soul (Condic and Condic 2005; Rao and Condic 2008; Condic 2003, 2008) seems to bear some oddities. For example, one can maintain a pregnant brain-dead woman for several weeks of gestation until the fetus is mature enough to be born. If a brain-dead body can exhibit visceral organ functioning for extended periods, grow and even gestate infants, it is somewhat arbitrary to claim that it is not an organism, an integrated whole (at least in some respect) (Laureys 2005). According to the "organization" criterion such a brain-dead woman should be considered a person. According to our criterion (and the clinical criteria for death) it can certainly be considered an organism (with a degree of "organization" comparable to that of a plant) but nonetheless dead (without animal life).

Similarly, it can happen that on the basis of the clinical criteria one declares death and blood chemistry suggests that the pituitary gland at the base of the brain is still functioning (Nature 2009). Though this activity may be considered a brain function, it is not a sign of capability for spontaneous motility, and therefore it does not demonstrate a person being alive in the sense of our criterion. In other words, as long as the clinical criteria ensure that a body declared brain dead will not recover *spontaneous* motility (like spontaneous breathing), they are criteria consistent with our philosophical premises.

Another oddity arises from "high cervical quadriplegic" patients with high-level spinal injuries (neck injuries) that leave them fully paralyzed and unable to breathe without technical assistance. Christopher Reeves is a famous example. Left alone, such patients die. They cannot be "cured," only maintained on life support. In many cases, they even need assistance to maintain a regular heart beat. If support is provided and the person does not die immediately, such patients remain fully conscious and capable of communication through strongly reduced motility mediated by the cranial nerves. However, their brains no longer control the function of their bodies. According to the "organization" criterion, the somatic integrative unity of such patients cannot be considered larger than that of a body declared to be brain dead. On this basis it is argued that the bodies of such patients are the equivalents of brain-dead bodies, especially if the patient also suffers from a transsection of the vagal nerve (Shewmon 2004). If nonetheless one considers them persons, this means that one is implicitly adopting the criterion of "neural activity" responsible for the reduced spontaneous motility (and not "organization") for ascertaining life.

Strictly speaking, the philosophical postulate that death means "the loss of the integration capacity of the organism" is not, in itself, sufficient to derive many of the conclusions derived from the current clinical criteria of death. One must additionally specify that the integration capacity referred to specifically involves the integrative capacity required to sustain the capacity for spontaneous motility.

In summary, one can distinguish four different possible philosophical positions regarding the status of organisms made of human cells:

1. Any organism made of human cells (including one that is brain dead) is a person.
2. There are human persons who are not organisms.
3. There are organisms made of human cells that possess vegetal and even specific animal features that are not persons. This seems to be the view behind the "delayed hominization" hypothesis and also the view behind the definition of death as neocortical or higher brain death (Laureys 2005).
4. There are organisms made of human cells (akin to a brain-dead body) possessing features of vegetal life which are not human beings or persons, but *any organism made of human cells and exhibiting specific animal features (like spontaneous motility in the persistent vegetative state) is animated by a spiritual soul and is a human being and a person*. This is the philosophical view we adopt, and which we also see as in most accord with the current clinical criteria for determining death.

5.5 Embryonic Development and Genomic Information

Embryonic development is comparable to a richly branched tree and shows several differentiation pathways. The first main branching is the differentiation of the inner cell mass (ICM) and the outer cell layer of the blastocyst or trophectoderm (TE), which after implantation gives rise to placental extraembryonic tissues (cyto- and syncitio-trophoblast). The second main branching consists in the differentiation of the ICM into epiblast and extraembryonic tissues (extraembryonic endoderm and yolk sac). The epiblast differentiates thereafter into the primitive ectoderm and amnionic ectoderm. The next main branching (during gastrulation) consists in the differentiation of the primitive ectoderm into *ectoderm* (external layer), *mesoderm* (middle layer), *endoderm* (internal layer), germ cells, and extraembryonic mesoderm. These three embryonic germ layers are the source of all tissues of the body (the germ cells originate from proximal epiblast cells that migrate to the extraembryonic mesoderm and here are committed to become primordial germ cells) (The National Institute of Health 2001a). In particular, the mesoderm differentiates mainly into bone marrow (blood), lymphatic tissue, connective tissues (bone, cartilage), vascular system, uro-genital system, skeletal, smooth and cardiac muscles; and the ectoderm into skin, brain, eyes, and pigment cells (The National Institute of Health 2001b). Spontaneous fetal motility can be observed in humans from the seventh week of pregnancy and said motility is considered a sign of neural activity (De Vries J.I.P. et al. 1982).

In sexual reproduction, each individual inherits one set of chromosomes from each parent – for each maternal chromosome, there is a corresponding paternal one. The chromosomes contain the genetic information. Normally there are two copies of each gene: one copy in a maternal chromosome and another copy in the corresponding paternal one. The maternal and paternal genomes are not exactly equivalent, but are endowed with different imprints or epigenetic modifications, regulating the expression of a number of genes during embryonic development. The imprints consist in chemical changes in the DNA or in the chromosomal proteins that may mark any gene without changing its DNA sequence and are transmitted through cell divisions (Loebel and Tam 2004). Embryonic development, then, largely depends on genomic (genetic and epigenetic) information.

In order for normal development to take place, the epigenetic inactivation of one gene from a gamete needs to be counterbalanced by the presence of the corresponding active gene from the other gamete. In certain cases, development fails if two copies of the same gene are active (one copy in the maternal-derived chromosome and another in the paternal-derived chromosome). Fully grown and secondary human oocytes contain two sets of chromosomes, one of which is normally discarded within 2 h after fertilization. Using a chemical treatment to prevent this, one produces diploid *parthenotes* with both sets of chromosomes coming from the mother (Rogers et al. 2004). In a mouse egg containing only chromosomes from fully grown oocytes, *H19* and several other genes are expressed twofold, whereas *Igf2* and several other genes remain inactive. These standard parthenotes appear to undergo the same initial changes as naturally fertilized eggs. They undergo cell cleavage for 4 or 5 days and some form blastocysts. Experiments in mice show that parthenotes are capable of undergoing early post-implantation development. However, a parthenogenetic ICM in combination with fertilization-derived TE can develop only until 12 days (birth in mice occurs after 19.5 days of gestation), reaching the heart beating stage but not the neural activity responsible for spontaneous fetal motility (Gardner et al. 1990).

Transferring the nucleus of a nongrowing (ng) primary oocyte (egg cell obtained from a new born mouse) into a fully grown oocyte results in a diploid cell, which contains two sets of chromosomes, both of maternal origin. Such artifacts are often called *ng-parthenotes*. We refer to them as *H19/Igf2* mutants too. In an ng-parthenote, several genes, such as *H19*, are twofold expressed and both copies of several other genes, such as *Igf2*, remain inactive, whereas some other genes are normally expressed. Such an *H19/Igf2* mutant mouse (or *ng-parthenote*) develops until 13.5 days. At day 12.5 it shows breathing and leg movements (Kono et al. 1996).

By knocking out the gene *H19* in the nucleus of the nongrowing oocyte before transfer, one produces an *Igf2-mutant* in which *H19* is monoallelically (single) expressed, whereas both *Igf2* copies remain inactive. Such a mouse mutant develops until 17.5 days (Kono et al. 2002). Finally, by knocking out the gene *H19* and activating the gene *Igf2* in the nucleus of the nongrowing oocyte before transfer one produces an epigenetic state closer to that of a fertilized egg, and indeed such a mouse embryo develops to birth around the 19th or 20th day (Kono et al. 2004).

5.6 The Embryo's *Proper Biological Potential* of Developing Neural Activity Is Not Determined by Implantation

Consider a body B exhibiting human architecture and spontaneous motility that derives through cell division from an embryo after implantation into a maternal uterus. Before implantation the embryonic layers ICM and TE interact with each other. After implantation the uterus interacts with TE and through it with ICM as well. The neural centers controlling the spontaneous movements of the body B derive through cell division exclusively from the ICM. No body exhibiting human architecture and spontaneous motility can derive through implantation of TE alone into a normal functioning uterus. If one separates the ICM from the TE, the ICM can give rise to a colony of ESCs, and the TE can degenerate into a tumor. From observations in mice one knows that the ICM in combination with a different healthy TE' produces a blastocyst ICM+TE' developing into a healthy adult B' (Gardner et al. 1990, control experiments). Again, the neural centers controlling the spontaneous movements of the body B' derive through cell division exclusively from ICM cell layer.

According to these observations the fact that the bodies B and B' exhibit spontaneous motility, and not only autonomous and reflex activities (like heart beating, blood circulation, kidney activity) depends exclusively on the neural centers deriving from cells of the ICM. That is, the ICM has the *proper biological potential* to develop neural activity. Thus, an ICM without TE is the same human being (has the same personal identity) as the body B, even if the ICM without a TE dies very quickly. This is similar to the way an embryo without implantation into a uterus is the same human being as B, even if the embryo without implantation dies very quickly. And further that B' is the same human being (has the same personal identity) as B and ICM. The TE can be replaced without changing the personal identity, just like the heart or skin of B. These conclusions are strengthened through the observations of epigenetic modifications in the embryo reported in Sect. 5.5. They show that neural activity is determined by *embryonic* information (the embryo's epigenetic state), just as a baby's sex or Down syndrome are determined by *embryonic* information (the Y chromosome and trisomy 21, respectively).

The interaction with the mother's organism certainly marks a biologically important step in embryonic development. One can even say that such interaction is necessary for the embryo to acquire complete developmental potential. Nonetheless the *proper biological potential* to develop neural activity is established well before implantation into the uterus. The fact that the neural centers derive from cells in the ICM and not in the TE, and the cells of the TE contribute only to build extraembryonic tissues like the placenta is not determined by implantation.

Consequently, a human embryo that would develop spontaneous fetal motility after implantation into a properly functioning uterus is animated by a spiritual soul and has the moral status of a human person. This means in particular that hypothetical *human ng-parthenotes* and *Igf2 mutants* deserve the moral status of persons. This also means that the interaction with the mother's organism is irrelevant for the moral status of the embryo, just as it is for sex determination or Down syndrome.

5.7 DIANA Insufficiencies as the Basis for Distinguishing Between Disabled Human Embryos and Nonembryos

The fact that, upon implantation in a properly functioning uterus, a human organism does not reach spontaneous fetal motility is not of itself sufficient to conclude that such an embryonic organism does not have the *proper biological potential* for neural activity and is not a person. Indeed, it could be a human person with a severe congenital defect who could have motility restored through an appropriate therapy. We need more precise rules for distinguishing between an organism that is a disabled human person and one that is a more or less organized assemblage of human tissues without personhood. This need is explicitly acknowledged by the Congregation for the Doctrine of Faith in the Instruction *Dignitas personae* (Congregation for the Doctrine of Faith 2008).

This goal remains important independently of the interest of deciding on parthenotes (Suarez et al. 2007) and cell entities derived through Altered Nuclear Transfer (ANT) (Hurlbut 2005a, b; Meissner and Jaenisch 2006) as possible alternative methods for obtaining pluripotent cells without destroying embryos. On the one hand, any consistent theory about the beginnings of human life should be able to provide general rules for distinguishing between embryos and nonembryos. On the other hand, recent achievements demonstrating that iPSCs can be used to produce viable mice through tetraploid complementation (Zhao et al.

2009; Boland et al. 2009; Kang et al. 2009) give rise to ethical concerns over use of such cells in the aim of producing a new technique in cloning. One may also doubt whether we are not reprogramming adult cells to the ICM stage. Accordingly, in the following paragraphs we try to devise criteria to distinguish between embryos and nonembryos.

First, let us consider the case of *androgenotes*. Androgenotes are diploid egg cells containing only the paternal genome produced from various anomalies that occurred during the fusion of gametes. Such cells arise sometimes in nature, if for instance there is a loss of the maternal chromosomes, together with a replication of the paternal chromosomes before cell division. Androgenotes carry faulty epigenetic information due to a lack of maternal genes: several genes become twofold expressed, whereas other genes remain inactive. Though exhibiting at the beginning an embryo-like development, androgenotes give rise to a blastocyst-like structure with a well-developed outer layer or TE but lacking the ICM. After implantation, androgenotes produce an entity called a "complete hydatidiform mole" in the uterus, having the appearance of a bunch of grapes, and whose cells remain undifferentiated. A type of tumor called chorion-epithelioma can arise from these entities. Androgenotes are affected by an epigenetic defect inhibiting the differentiation of the proper embryonic tissues and inducing only the differentiation of extraembryonic ones. Since the neural centers of any body do not derive from extraembryonic tissues, androgenotes lack the *proper biological potential* for unfolding neural activity or, in other words, the anomaly they carry *directly* excludes the capacity for this activity. Such an anomaly, according to our philosophical main assumption, makes an organism intrinsically incapable of being animated by a human soul.

The reverse case would be an epigenetic anomaly leading to a defective TE but a well-formed ICM, so that if one would inject this ICM into a healthy TE, the reconstituted blastocyst would be able to implant and develop to birth. Such an epigenetic anomaly does not *directly* frustrate the development of neural activity. Although the defective TE would prevent the implantation, and thereby provoke the organism to die, its ICM has the *proper biological potential* for the emergence of neural activity. Said epigenetic anomaly could be compared to an intervention that thwarts the implantation of a normal embryo into the uterus. We consider the organism carrying such a genomic anomaly to be a disabled embryo and a human person.

Consider now the case of *standard parthenotes* referred to earlier. Evidence in mouse models by Richard Gardner and colleagues (1990) show that by combining a parthenogenetic primitive ectoderm with a fertilization-derived TE one can obtain parthenotes that reach the stage of having a vigorous heart beat and visceral yolk sac circulation, though they do not appear to go beyond this point. This means that standard parthenotes fail to reach fetal motility not because of a defective placenta or heart, but because of defects in the development of the brain itself. These results support the conclusion that parthenotes are organisms capable of developing heart beat and blood circulation but incapable of unfolding the neural activity required for spontaneous fetal motility. We assume that from a moral point of view such organisms share the same status as brain-dead adults (i.e., after irreversible breakdown of the cortex and the brain stem), and should not be considered persons. Like brain-dead adults (see Sect. 5.4), standard parthenotes can be considered organisms sharing features of vegetal life ("vegetal soul"), i.e., biological systems with a degree of organization comparable to that of plants. Standard parthenotes do not share the specific observable feature of animal life, i.e., the *proper biological potential* of spontaneous motility. In more philosophical terms, one could say that a parthenote lacks the biological conditions required for animation by a human spiritual soul in much the same way as does a brain-dead adult.

The evidence that parthenogenetic ICM cells participate in normal embryonic development of chimeric animals following injection into blastocysts, and appear most likely to be incorporated into the brain (Strain et al. 1995; Boediono et al. 1999), does not speak against the equivalence between parthenotes and brain-dead adults. As loss of brain function in a brain-dead body does not lead to the death of all cells in the brain, the incapacity for neural activity in parthenotes does not necessarily exclude the appearance of brain cells.

Consider further the case of an epigenetic defect that after gastrulation (formation of the three embryonic tissue layers in the early embryo) inhibits the healthy differentiation of the cardiac muscle cells. As a consequence, the organism dies at mid-gestation, but not because of a direct inhibition of the emergence of neural activity. The genomic configuration does not directly, but indirectly, prevents neural activity. The knockout of three *Id* genes may be an example of such an alteration. The observations in mice models suggest that a defective heart is the primary cause of the resulting mutants' death at about day 11 of gestation. This mid-gestation death can be avoided by the injection of a low number of wild-type ESCs into mutant blastocysts. As a result, the myocardial markers altered in *Id* mutant cells are restored to normal throughout the chimeric myocardium (Fraidenraich et al. 2004). We think that such a defect is easily comparable to a lethal heart injury, and their restoration to heart transplantation. Accordingly, we think that in this case the comparison with a brain-dead body does not hold, and assume that such an organism shares the same status as an adult with a healthy brain but with lethal injuries to other vital organs. Therefore, from a moral point of view, the hypothetical human *Id* mutants should be considered disabled embryos and deserve the moral status of persons.

Notice that in case of the *Id*-mutants referred to earlier, the blastocysts, when placed into a properly functioning uterus environment, would behave the same way as standard parthenotes, that is, they would soon die and produce disorganized tissue. Nevertheless, it is not reasonable to conclude that the corresponding embryo-like entities were nonembryos.

Let us finally look at the case of $Cdx2^{-/-}$ cell entities or mutants obtained through deletion or inactivation of the Cdx2 gene. Such an alteration, even from the 2-cell stage, affects the TE lineage but not the ICM lineage. Thus, we consider that inactivating Cdx2 does not impair the *proper biological potential* for neural activity and leads to the failure of brain development only *indirectly*. In this sense, $Cdx2^{-/-}$ mutants are rather comparable to the *Id*-mutants reported earlier and cannot be compared to brain-dead organisms. A "brain dead" body is a good example of an organism in which all organs but the brain, function well. By contrast, as far as we know, there is no experimental basis to state that the inactivation of the gene Cdx2 *directly* precludes the development of neural activity. We stress that by questioning the production of $Cdx2^{-/-}$ cell entities we do not question the method of ANT (Hurlbut 2005a; Meissner and Jaenisch 2006) as such, but only the implementation of ANT through inactivation of the Cdx2 gene.

We use the abbreviation *DIANA* (directly inhibiting the appearance of neural activity) to denote genomic configurations or epigenetic states that are incapable of supporting the embryonic development on the pathway leading *directly* from the zygote to the emergence of the neural centers that trigger fetal motility. The pathway is:

zygote → ICM → epiblast → primitive ectoderm → ectoderm → brain cells → neural activity

Insufficiencies frustrating the development on side branches that do not lead to differentiation of brain cells and appearance of neural activity are *not DIANA* alterations, even if they may be lethal and lead *indirectly* to the failure of brain development, as a consequence for instance of a defective heart or placenta.

We sum up our view in the following three principles:

1. Any anomaly of the developmental program, including one or several *DIANA* insufficiencies, renders an organism inadequate to be animated by a human soul.
2. In the absence of any *DIANA* insufficiency one must conclude that a human soul is present and consider the organism a person with a possibly severe congenital defect.
3. If there is doubt about whether an organism exhibits a *DIANA* insufficiency, one cannot exclude the presence of a human soul.

Against this view one could object that a *DIANA* insufficiency may be *reversible* (for instance, one could remove it through appropriate epigenetic modifications of the cells at an early developmental stage), whereas brain death follows from an *irreversible* damage. However, this objection does not invalidate our principles. We assume that a human soul is present, if the cell or organism owns the intrinsic potentiality to unfold the neural activity responsible for spontaneous movements. In the case of brain death we have a body, which earlier exhibited such movements. To declare the person dead we must be sure that his or her body no longer has the potential for spontaneous motility, and for this reason we test whether it has irreversibly lost the capacity of performing such movements. By contrast, a human cell that has a *DIANA* genomic configuration from the very beginning will never develop into a body exhibiting spontaneous motility, and we assume that it is not a person. If one removes the insufficiency at a later moment and produces an organism capable of reaching spontaneous fetal motility, then we assume that a new person appears at the very moment the alteration disappears just as we assume that a new person appears after fertilization.

5.8 Can the ICM Be Considered Equivalent to a Clump of ESCs?

The main objection against the *proper biological potential* and DIANA criteria is based on the experimental generation of fully ESCs-derived mice, which has been done either using complementation with a tetraploid embryo (Rossant and Spence 1998), or laser-assisted injection of ESCs into eight cell-stage diploid embryos (Poueymirou et al. 2007).

Suppose one accepts our criterion for distinguishing between embryos and nonembryos based on their *proper biological potential* to develop neural activity. Since the neural centers of fully ESCs-derived mice consist of cells originating from the ESCs, one could argue that a colony of ESCs possesses itself the proper potential for neural activity and shares embryo status and this seems to be a rather awkward conclusion.

Suppose one claims that the ICM without a TE is equivalent to a clump of ESCs. That is, both the ICM and the ESCs acquire *proper biological potential* only through the interaction with TE. If this were the case, cell entities carrying genomic alterations that prevent the development of the TE lineage from the very beginning should be considered a colony of ESCs and not a disabled embryo, in contradiction to the DIANA criterion we offered earlier.

Against these objections we reply that generation of fully ESCs-derived mice has always been done using combination of ESCs with embryos (morulae or tetraploids) *without removing the ICM from the host*. In either case one cannot exclude the possibility that the ESCs experience changes in their gene expression pattern equivalent to what they experience when injected into a diploid blastocyst with ICM.

To solve this dilemma, it would be important to perform the following experiment in a mouse model: after the extraction of the ICM from a fertilization-derived blastocyst, one injects a colony of ESCs into the remaining healthy TE (Actually this experiment has been done but the results have not been published so far; for a more precise description see Suarez 2011, the chapter 11 in this book). Our prediction is that the so-reconstituted blastocyst will experience the same fate as the blastocysts reconstituted using parthenogenetic primitive ectoderm in combination with fertilization-derived TE (Gardner et al. 1990), i.e., they may possibly reach heart beating and blood circulation, but will not unfold spontaneous fetal motility. It is well known as well that the ICM cells are heterogeneous and only a subpopulation of them gives rise to ESCs (Zwaka and Thomson 2005). Observations on developmental markers show also that ICM and ESCs cannot be considered equivalent in spite of the fact that ESCs remain undifferentiated (Wongtawan et al. 2011). There is additionally evidence that in chimeric mouse blastocysts the tetraploid cells are depleted in the epiblast region of the late blastocyst but not completely excluded until later (MacKay and West 2005). These results seem to speak in favor of our prediction that using ESCs in combination with diploid TE will not have the capability to develop neural activity. If our prediction is confirmed by experiment, then it would be clearly demonstrated that a colony of ESCs has a developmental potentiality similar to that of the standard parthenotes and cannot be considered equivalent to an ICM. Thus, the DIANA criterion would be proved right by experiment.

By contrast, if our prediction fails, one would be led to the conclusion that the ESCs colony acquires the capability of developing neural activity through the interaction with the TE. If this holds, alterations preventing the development of the TE lineage would not produce disabled embryos but simply clumps of ESCs. Nonetheless, one would have to face two challenges: (1) demonstrating that the modifications the TE induces into the ESCs and those the implantation induces into the ICM cells are essentially different from each other and (2) determining when, after implantation, the extraembryonic layers cease being necessary for the personal identity and *moral* status of the ICM-derived body. Without achieving (1) and (2) one could not invalidate the argument that the embryo requires the interaction with the maternal organism (or an equivalent growth culture) in order to acquire "full developmental potentiality and moral status."

In any case, the proposed experiment would allow us to ascertain whether aggregation with a tetraploid embryo or a diploid eight cell-stage induces changes into the ESCs, which are essentially different from the changes induced by the mere TE. And such an experiment is valuable not only from an ethical but also a scientific point of view.

5.9 Can Reprogramming Adult Somatic Cells Become a New Form of Cloning?

The experimental generation of fully iPSCs-derived mice using complementation with a tetraploid embryo has been considered the proof of complete equivalence between ESCs and iPSCs (see section 5.7). However, the possibility of obtaining fully iPSCs-derived animals raises a new concern.

Apparently, the contact with the tetraploid embryo reprograms *in vivo* the ESCs and iPSCs to a state equivalent to that of an ICM interacting with the TE. Consequently, nothing speaks in principle against the possibility of reprogramming *in vitro* somatic cells to iPSCs equivalent to an ICM and, therefore, sharing the status of an embryo. From an ethical point of view it is important to exclude this new form of cloning.

Achieving this requires an experiment similar to the one proposed in the preceding section. That is, inject a colony of iPSCs into a healthy TE without ICM. If the so-reconstituted blastocyst iPSCs+TE fails to develop, one can be sure that the iPSCs are not equivalent to an ICM. Suppose by contrast that the reconstituted blastocyst iPSCs+TE develops to birth while the blastocyst ESCs+TE does not. This would mean that the reprogramming has gone too far and produced a cell entity sharing the status of an embryo (see Suarez 2011, the chapter 11 in this book).

5.10 Conclusions

We have argued that personhood, or the presence of a rational soul, can be ascertained in a human organism by observing the presence of the *proper biological potential* to unfold the neural activity responsible for controlling spontaneous motility. Additionally, we have shown that the embryo's proper potential of developing neural activity is not determined by implantation. Hence, we are led to conclude that a human embryo in culture is animated by a human spiritual soul.

Our inquiry also shows the need, and benefit, of fine-tuning the conventional definition of death as the "loss of the integrative capacity specific to an animal organism." In our view, a brain-dead animal organism is declared dead on the basis of the irreversible loss of the proper potential for spontaneous motility in spite of the fact that it may still exhibit a certain degree of "integration" and be considered a "living organism" to some extent. This means that our philosophical categories support the current clinical criteria for defining death.

We have further introduced the concept of insufficiencies that directly inhibit the appearance of neural activity (DIANA insufficiencies). We have shown that DIANA insufficiencies provide sufficient criteria for distinguishing between disabled embryos and nonembryos. This task remains relevant independently of the fact that iPSCs techniques have superseded the possible use of parthenotes or ANT for deriving human pluripotent cells without destroying embryos. Indeed, the iPSCs techniques themselves may become improved to the point of reprogramming adult cells to an ICM or morula stage.

Finally, we have argued that the criteria of *proper biological potential* and DIANA cannot be objected to on the basis of experimental generation of fully ESCs-derived mice using aggregation of ESCs with diploid or tetraploid morulae, for such morulae remain capable of producing an ICM. To settle this issue it is necessary to perform a new experiment (using well-known techniques along the lines outlined earlier) in which ESCs are aggregated into a blastocyst without an ICM.

References

Aquinas Thomas (CTh) Compendium theologiae, ch. XCII. http://dhspriory.org/thomas/Compendium.htm. Cited 14 March 2011

Aristotle (MA) On the motions of animals, Part 11. The Internet Classics Archive http://classics.mit.edu/Aristotle/motion_animals.html. Cited 26 Feb 2011

Aristotle (NE) Nicomachean ethics, Book III, 1. The Internet classics archive http://classics.mit.edu/Aristotle/nicomachaen.3.iii.html. Cited 14 March 2011

Ashley B, Moraczewski A (2001) Cloning, Aquinas and the embryonic person. Natl Cathol Bioeth Q 1:189–201

Boediono A, Suzuki T, Li LY, Godke RA (1999) Offspring born from chimeras reconstructed from parthenogenetic and in vitro fertilized bovine embryos. Mol Reprod Dev 53:159–170

Boland MJ, Hazen JL, Nazor KL, Rodriguez AR, Gifford W, Martin G, Kupriyanov S, Baldwin KK (2009) Adult mice generated from induced pluripotent stem cells. Nature 461:91–94

Check E (2006) Simple recipe gives adult cells embryonic powers. Nature 442:11
Condic ML (2003) Life: defining the beginning by the end. First Things 133:50–54
Condic ML (2008) When does human life begin? A scientific perspective. The Westchester Institute for Ethics & the Human Person. White Paper 1:1–18
Condic ML, Condic SB (2005) Defining organisms by organization. Natl Cathol Bioeth Q 5:339
Congregation for the Doctrine of Faith (2008) Instruction *Dignitas personae*, p. 18 and footnote 49
De Vries JIP, Visser GHA, Prechtl HFR (1982) The emergence of fetal behaviour: I Qualitative aspects. Early Hum Dev 7:301–322
Denker HW (2009) Ethical concerns over use of new cloning technique in humans. Nature 461:341
Fraidenraich D, Stillwell E, Romero E, Wilkes D, Manova K, Basson CT, Benezra R (2004) Rescue of cardiac defects in id knockout embryos by injection of embryonic Stem Cells. Science 306:247–252
Gardner RL, Barton SC, Surani MA (1990) Use of triple tissue blastocyst reconstitution to study the development of diploid parthenogenetic primitive ectoderm in combination with fertilization-derived trophectoderm and primitive endoderm. Genet Res 56:209–222
González F, Boué S, Izpisúa Belmonte JC (2011) Methods for making induced pluripotent stem cells: reprogramming á la carte. Nature Rev. Genet. 12:231–242
Huarte J, Suarez A (2004) On the status of parthenotes. Defining the developmental potentiality of a human embryo. Natl Cathol Bioeth Q 4:535–550
Hurlbut WB (2005a) Altered nuclear transfer as a morally acceptable means for the procurement of human embryonic stem cells. Perspect Biol Med 48:211–28
Hurlbut WB (2005b) Correspondence: altered nuclear transfer. New Engl J Med 352:1153
Jones DA (2004) The soul of the embryo: an enquiry into the status of the human embryo in the Christian tradition. Continuum, London, Ch. 8 (The timing of ensoulment)
Kang L, Wang J, Zhang Y, Kou Z, Gao S (2009) iPS can support full-term development of tetraploid blastocyst-complemented embryos. Cell Stem Cell 5:135–138
Kono T, Obata Y, Yoshimzu T, Nakahara T, Carroll J (1996) Epigenetic modifications during oocyte growth correlates with extended parthenogenetic development in the mouse. Nat Genet 13:91–94
Kono T, Sotomaru Y, Katsuzawa Y, Dandolo L (2002) Mouse parthenogenetic embryos with monoallelic H19 expression can develop to day 17.5 of gestation. Dev Biol 243:294–300
Kono T, Obata Y, Wu Q, Niwa K, Ono Y, Yamamoto Y, Park ES, Seo JS, Ogawa H (2004) Birth of parthenogenetic mice that can develop to adulthood. Nature 428:860–864
Laureys S (2005) Science and society: death, unconsciousness and the brain. Nat Rev Neurosci 6:899–909
Libet B (1985) Unconscious cerebral initiative and the role of conscious will in voluntary action. Behav Brain Sci 8:529–566
Libet B (1992) Voluntary acts and readiness potentials. Electroencephalogr Clin Neurophysiol 82:85–86
Libet B (2002) The timing of mental events: Libet's experimental findings and their implications. Conscious Cogn 11:291–299
Loebel DAF, Tam PPL (2004) Genomic imprinting. Mice without a father. Nature 428:809–811
MacKay GE, West JD (2005) Fate of tetraploid cells in 4n↔2n chimeric mouse blastocysts. Mech Dev 122:1266–1281
Meissner A, Jaenisch R (2006) Generation of nuclear transfer-derived pluripotent ES cells from cloned Cdx2-deficient blastocysts. Nature 439:212–215
Nature (2008) Life after SuperBabe (Editorial), and making babies: the next 30 years (Special Report by H. Pearson). Nature 454: 253, 260–262
Nature (2009) Delimiting death (Editorial). Nature 461:570
Nüsslein-Volhard Ch (2004) Das Werden des Lebens. Verlag CH Beck, Munich, pp 189–191
Plath K, Lowry WE (2011) Progress in understanding reprogramming to the induced pluripotent state. Nature Rev. Genet. 12:253–265

Poueymirou WT, Auerbach W, Frendewey D, Hickey JF, Escaravage JM, Esau L, Doré AT, Stevens S, Adams NC, Dominguez MG, Gale NW, Yancopoulos GD, DeChiara TM, Valenzuela DM (2007) F0 generation mice fully derived from gene-targeted embryonic stem cells allowing immediate phenotypic analyses. Nat Biotechnol 25:91–9

Rao M, Condic ML (2008) Alternative sources of pluripotent stem cells: scientific solutions to an ethical dilemma. Stem Cells Dev 17:1–10

Rizzolatti G, Craighero L (2004) The mirror-neuron system. Annu Rev Neurosci 27:169–192

Rodríguez-Luño A (2006) Il dibattito sull'identità e lo statuto dell'embrione umano. Etica e politica, Bioetica. http://www.eticaepolitica.net/bioetica.htm Cited 26 Feb 2011

Rogers NT, Hobson E, Pickering S, Lai FA, Braude P, Swann K (2004) Phospholipase Cz causes Ca2+ oscillations and parthenogenetic activation of human oocytes. Reproduction 128: 697–702

Rossant J, Spence A (1998) Chimeras and mosaics in mouse mutant analysis. Trends Genet 14:358–363

Schöler H (2007) Ein Embryo ist kein Fötus und eine Zelle kein Mensch. Frankfurter Allgemeine Zeitung, 11 September

Shewmon A (2004) The 'critical organ' for the organism as a whole. In: Machado C, Shewmon A (eds) Brain death and disorders of consciousness. Kluwer, New York, pp 26–35

Strain L, Warner JP, Johnston T, Bonthron DT (1995) A human parthenogenetic chimera. Nat Genet 11:164–169

Suarez A (1993) Sono l'embrione umano, il bambino con anencefalia ed il paziente in stato vegetativo persistente persone umane? Una dimostrazione razionale a partire dai movimenti spontanei. Acta Philosophica 2:105–125

Suarez A, Lang M, Huarte J (2007) DIANA Anomalies. Criteria for generating human pluripotent stem cells without embryos. Natl Cathol Bioeth Q 7:315–335

Suarez A (2011) Is this cell entity a human being? Neural activity, spiritual soul, and the status of induced pluripotent stem cells (iPSCs). In: Suarez A, Huarte J (eds) Is this cell a human being? Springer, Berlin, pp 171–192

Takabashi K, Yamanaka S (2006) Induction of pluripotent stem cells from mouse embryonic and adult fibroblast cultures by defined factors. Cell 126:663–676

The National Institute of Health (2001a) Stem cell report, Appendix A: Early development. http://stemcells.nih.gov/info/scireport/appendixa.asp. Cited 26 Feb 2011

The National Institute of Health (2001b) Stem cell report, Chapter 1: The stem cell http://stemcells.nih.gov/info/scireport/chapter1.asp. Cited 26 Feb 2011

Wongtawan T, Taylor JE, Lawson KA, Wilmut I, Pennings S (2011) Histone H4K20me3 and HP1 are late heterochromatin markers in development, but present in undifferentiated embryonic stem cells. J. Cell Sci 124:1878–1890

Zhao XY, Li W, Lv Z, Liu L, Tong M, Hai T, Hao J, Guo CL, Ma QW, Wang L, Zeng F, Zhou Q (2009) iPS cells produce viable mice through tetraploid complementation. Nature 461:86–90

Zwaka TP, Thomson JA (2005) A germ cell origin of embryonic stem cells? Development 132:227–233

Chapter 6
Distinguishing Embryos from Non-embryos

Patrick Lee

Abstract This chapter considers several techniques recently developed to obtain human pluripotent stem cells, without, apparently, disaggregating human embryos: reprogramming somatic cells (to obtain induced pluripotent stem cells, or iPS cells), altered nuclear transfer combined with oocyte-assisted reprogramming, and parthenogenesis (naturally occurring or chemically activated). On what basis can one determine whether a product of a reproductive technique is a human embryo or not? This chapter proposes a criterion: does the entity produced have the genetic–epigenetic state, and overall organization, such that it will develop itself to the mature stage of a human organism (an organism with a brain that can provide experience suitable to be the substrate of conceptual thought), provided a suitable environment and nutrition? This chapter briefly defends that criterion and applies it to various types of biological entities.

Keywords Embryo • Formal cause • Human organism • iPS cells • Material cause • Mechanism • Parthenogenesis • Tetraploid complementation

6.1 Introduction

Recently, several techniques have been developed to obtain human pluripotent stem cells, without, apparently, disaggregating human embryos. These techniques include reprogramming somatic cells (producing induced pluripotent stem cells, or iPS cells), altered nuclear transfer combined with oocyte-assisted reprogramming (ANT-OAR), and parthenogenesis (e.g., by chemically activating the female pronuclei in oocytes). These proposed techniques raise the following question: how can one determine that a product of a reproductive technique is a cell or

P. Lee
Franciscan University of Steubenville, Steubenville, OH, USA
e-mail: plee@franciscan.edu; plee512@gmail.com

A. Suarez and J. Huarte (eds.), *Is this Cell a Human Being?*,
DOI 10.1007/978-3-642-20772-3_6, © Springer-Verlag Berlin Heidelberg 2011

a group of cells that is *not* a human embryo? Is it possible that such techniques might produce disabled embryos rather than non-embryos?

From the ethical angle, we are primarily interested in entities that qualify as subjects of rights. I presuppose here that every *rational being* – that is, an individual possessing a rational nature – is a subject of rights. [There are several cogent defenses of this position. For example, see Lee and George (2008a)]. Every human being has a rational nature, even though it may take him or her several months to actualize that nature, that is, to perform the specific operations to which that nature orients him or her (Lee 2004; Lee and George 2008b; Oderberg 2000). Thus, an individual need not be able to perform rational actions *right now*, such as self-awareness and deliberate choice, in order to be a subject of rights. It is enough if he or she has a nature orienting him or her to those types of operations. Every human being, that is, every human organism, has a rational nature. Every human being has a structure orienting him or her to the mature stage at which he or she will be able to perform such actions. (I hold that human beings do possess rational souls. But one does not determine whether an entity is a human being by first determining whether it has a rational soul. Rather, one first finds evidence that an entity is a human being and from that one concludes that he or she must have a rational soul. In my judgment, conceptual thought, acts of will, and reflexive self-awareness are immaterial actions, not themselves brain processes. But in human beings those actions *depend on* previous brain processes sufficiently complex to produce sense presentations suitable to be the source and specifying cause of basic, first-level conceptual thought. So, every human being must either have a brain or have the capacity to develop a brain. Thus, while I hold that a human being has a rational soul, which survives death, and which cannot emerge from lower material forces – these are not points that must be presupposed for examining these issues.) It follows that just as a human being who is in a coma is a subject of rights, though he cannot presently exercise his basic natural capacities for rationality and deliberate choices, in the same way, a human embryo or fetus is a subject of rights, though it may take him or her several months to actualize his or her specifically human basic capacities.

At what point does a human embryo come to be? In the normal case of reproduction, the answer is that, when at fertilization the sperm joins the ovum, the sperm and the ovum cease to be, and generate a new entity, which is a distinct new organism that has the developmental trajectory toward, or the active disposition to develop itself to, the mature stage of a human being. Put differently: when a human sperm has joined a human ovum, and generates a new and distinct organism, there now exists an organism whose genetic–epigenetic state, and overall organization, is such that it will develop itself to the mature stage of a human organism, provided a suitable environment and nutrition.

In the last paragraph I stated the basic criterion for being a human organism in two ways, or rather, I stated two aspects of that criterion. One aspect emphasizes the formal or final cause: a new human being is generated when a new organism that has the active disposition to develop itself to the mature stage of a human organism is produced. The evidence for the existence of such an entity would be the developmental trajectory of such a being when placed in the appropriate environment and provided sufficient nutrition. As the scholastics put it: *agere sequitur esse*.

[Father Nicanor Austriaco, OP, emphasizes this point in Austriaco (2005)]. Stating the issue in this way explicitly recognizes that we are speaking of the organism as a living, whole entity, not a mere aggregate of smaller entities [on this see Oderberg (2007, especially 65–71) and Connell (1988)].

The second aspect of the criterion is that a new human organism exists when there is a new living entity with the genetic–epigenetic state that provides the cascaded information for the development of this entity to the mature stage of a human organism. This states the criterion from the standpoint of the material cause. (By material cause is meant that out of which a thing comes to be, its ingredients. By formal cause is meant the actualization or specification of these ingredients to make them an individual of this fundamental kind rather than of another kind.) Of course, these two aspects of the criterion (formal cause and material cause) are metaphysically interdependent; each implies the other. One can use the criterion in either of its aspects to distinguish one type of cell from another, for example, a skin cell from a one-celled embryo, a zygote. The human zygote has an active disposition to develop itself to the mature stage of a human organism; the skin cell does not. The human zygote has the genetic and epigenetic state orienting it to develop to the mature stage of a human being, but the skin cell does not. If one found a tiny cell in a Petri dish in one's lab, one could determine with certainty whether it is a human embryo as opposed to a skin cell. In most cases, by this criterion one could also tell whether it was a severely disabled embryo or a non-embryo of some type.

Both a skin cell and a one-celled embryo (zygote) have the whole human genome (DNA sequence in its nucleus), but the zygote is a whole (though immature) human organism, while the skin cell has a nature that makes it suitable to function only as a small part of a human organism. Thus, the mark of a human organism at its early stages of development is not just the genetic structure in its cells, but the total epigenetic state of its cells – how the DNA within the nuclei of these cells is modified to determine the order and sequence of differential gene expression – the cytoplasmic factors within the cells, and (at the multicellular stages) how these cells are arranged.

The process of fertilization sometimes goes seriously awry and can generate an entity (or a group of cells) that is human in the sense that its cells have nuclei with the human genome, but this entity lacks the constitution that would orient it to developing itself to the mature stage of a human. Such an entity is not a human organism. The criteria mentioned above can distinguish between such products of incomplete fertilization and human embryos. What follows are some applications.

6.2 Various Type of Conceptus

6.2.1 Complete Hydatidiform Moles

A complete hydatidiform mole (CHM) is a conceptus that contains placental tissues and contains only genes of paternal origin, though it is diploid. It is formed in one of two ways: (a) two spermatozoa fertilize an oocyte that has lost its nucleus and the two male pronuclei then fuse to form a diploid nucleus, or (b) a single sperm fertilizes an oocyte that lacks its nucleus and the chromosomes in the sperm's nucleus duplicate without separating and form a diploid nucleus. The CHM cells lack the epigenetic state that orients them to the organized self-development toward the mature stage of their species. From the beginning, CHMs lack the organization necessary to divide into trophoblast and inner cell mass – the first differentiation in the embryo's development. The various cells in this growth at no point act together as parts of a whole – indicating that they *are* not parts of a whole. Although the cells contain the human genome, the epigenetic state of these cells is not such as to dispose them as a unit to develop to the mature stage of any organism. Thus, given the criterion stated above, it follows that CHMs are not at any stage human organisms.

6.2.2 Partial Hydatidiform Moles

A partial hydatidiform mole results from the fertilization of a normal oocyte by two spermatozoa to form a conceptus with three sets of chromosomes (not the normal diploid). Some of these entities have differentiated into trophoblast and inner cell mass (the precursors, respectively, of the placental organs and the rest of the embryo) and it also seems possible that they possessed an orientation toward developing a brain of the human sort, but also possessed a defect (in a part of their genetic makeup distinct from those parts providing an orientation to the development of a brain and nervous system) that precluded their survival. So, a partial hydatidiform mole may contain, or may have contained, an immature human organism within it. [A partial hydatidiform mole usually also contains extraneous cells attached to it.]

6.2.3 Parthenotes Occurring in Nature

A parthenote is in some ways comparable to a CHM. A CHM contains only paternally derived chromosomes; a parthenote (if not manipulated) contains only maternally derived chromosomes. (In fact some authors refer to CHM's as androgenotes) A parthenote is the product of parthenogenesis, that is, the activation of an oocyte without the fusion of a spermatozoan. In some nonmammalian species, parthenogenesis is a normal way of producing live offspring. In mammals, parthenogenesis is abnormal, but does occur rarely; and such parthenotes develop for a short time in an embryonic manner (by cell cleavage, and forming a cohesive group of cells). But, when produced without artificial intervention, the parthenote does not survive past the eight-cell stage (Brevini and Gandolfi 2008; Huarte and Suarez 2004). Like CHMs, such parthenotes from the beginning lack even the capacity to differentiate into inner cell mass and trophoblast, the first differentiation occurring in embryonic development. Thus, mammalian parthenotes occurring in nature are not human embryos.

6.2.4 Genetically Manipulated Parthenotes

Recently, techniques have been developed to *induce* parthenogenesis in mammals, including mice, rabbits, cows, pigs, goats, and primates. At the same time, the process has been modified so that these parthenotes will develop further than the first few cell divisions. Induced parthenotes require further analysis.

In normal fertilization, both paternally derived and maternally derived genes function in orchestrated ways, each silencing certain genes of the opposite type and (at appropriate times) activating transcription of certain genes in its own nucleus. This complementary silencing and expression of different genes is called *genomic imprinting*. The lack of genomic imprinting in CHMs and naturally occurring parthenotes is the central reason why they are not whole mammalian organisms. However, when parthenogenesis is induced, the genetic expression in these cells is sometimes also altered in order to compensate for the lack of genomic imprinting. This raises the question of whether such a technique might produce an embryo. A process of altered parthenogenesis could in effect become a type of cloning, rather than the activation of an oocyte to develop in an embryo-like manner.

There are three types of altered or manipulated parthenogenesis – all have been attained with mice and some also with humans. I will leave the discussion of the scientific details of the different types of parthenogenesis to others, but suffice it to say that some techniques merely coax the oocyte to maintain a diploid state (since a matured, or secondary, oocyte has two sets of chromosomes before it ejects half of them in what is called the second polar body), without supplying information that would normally come from the sperm. (The procedure consists in exposure of oocytes to a calcium ionophore followed by a protein synthesis inhibitor – to mimic changes normally induced by the sperm in fertilization [see Huarte and Suarez (2004) and references there; also Paffoni et al. (2007)]). These parthenotes only develop to the blastula stage, and it is clear that they lack the intrinsic potential to develop a basic body plan and, a fortiori, from the beginning lack the intrinsic potential to develop a neural system. Hence, these are not human organisms.

By a *second* type of parthenogenesis, which has been applied to mice, it seems possible that one could produce a human parthenote that could develop to the stage where the heart forms and some circulation begins, as well as some neural activity. In this procedure, an immature oocyte is combined with a mature one (the immature oocyte is called a nongrowing oocyte or ng oocyte, and so these are referred to as ng parthenotes). By a *third* procedure, a human parthenote might be produced, by combining an immature oocyte with a mature one, and at the same time altering a gene (silencing the H19 gene) in one of the oocytes before fusion. By this procedure, the entity produced develops even further than the entities produced by the second procedure but still not to birth (17.5 days in mice, where birth is typically at day 19.5). And by a *fourth* procedure one might create a human parthenote that develops to birth. (This is like the third process, but in addition, another gene that would otherwise be abnormally silent in both oocytes (Igf2) is activated.) [See Huarte and Suarez (2004), also Paffoni et al. (2007).]

Huarte and Suarez have argued that the first procedure does not produce embryos, but the second, third, and fourth procedures probably do. They argue that the minimally sufficient potential for development needed to be a human embryo is the capacity to develop spontaneous motility, since such a capacity is a prerequisite for voluntary action, and such deliberative action is the operation distinctive of the human kind (Huarte and Suarez 2004; Suarez et al. 2007).

If the scientific argument they advance is correct, then I agree with their overall, philosophical conclusion – philosophical, because it partly depends on a philosophical premise. Whether a developing entity is a human organism must be answered by determining (if possible) whether it has the intrinsic active potential to develop to the stage where it can (if not impeded) perform operations specific to human organisms. Human organisms are rational animals. A rational animal is one that has at least a radical capacity to perform rational functions. An organism has a *radical* capacity or potentiality for a function if has within itself a material constitution that disposes it, given a suitable environment, to develop sufficiently to perform that function, even if it cannot right now perform that function. The rational operations that differentiate human beings from other animals include such functions as conceptual thought, reflexive self-awareness, reasoning, and making deliberate choices. Rational functions presuppose sensation, and in mammalian organisms such as humans, sensation presupposes a functioning brain. It follows that an entity is a human being only if it has a radical capacity for sentience, and thus, only if it either has a brain or has the intrinsic capacity to develop a brain [see Grisez and Lee (2011)]. Thus, parthenotes that lack the capacity to develop a brain are not human organisms, while parthenotes that do have such a capacity *are* human organisms; so the conclusions of Huarte and Suarez about the different types of parthenotes are correct.

6.2.5 Products of ANT-OAR

Altered nuclear transfer (ANT), joined with oocyte-assisted reproduction (OAR), is a modification of somatic cell nuclear transfer (SCNT, or cloning) in order to produce a mass of cells that do not constitute a human organism, but from which pluripotent stem cells can be cultured. In the ANT-OAR process, the nucleus of a somatic cell is joined to an enucleated oocyte (as in cloning), but before that, certain genes in the nucleus of the somatic cell are silenced, and certain genes in the oocyte are over-expressed, so that the product of the fusion of these entities is not a human organism.

One proposal is to knock out the Cdx-2 gene in the nucleus of the somatic cell before joining it to the enucleated oocyte and at the same time overexpress the nanog gene in the enucleated oocyte. Transcription of the Cdx-2 gene and the delay of transcription of the nanog gene are necessary for the differentiation of the cells into inner cell mass (the precursors of the permanent organs in the embryo) and trophoblast or trophectoderm (the cells that are the precursors to the placenta, chorion, and other temporary organs of the embryo). But this differentiation is the first step made by the human organism as a whole in its progressive differentiation and self-development to a highly complex body. So, if the product of ANT-OAR from the beginning lacks this gene (that is, it has been repressed or inactivated in some way), and has the nanog gene from the ooctye expressed, then it lacks constituents and structure necessary for making the first step in human development. To possess such an epigenetic state at this stage is not just to lack a part in an otherwise intelligible whole; rather, it is to lack a structure suited to the development toward a more complex multicellular organism. Thus, if the scientific basis of the argument is correct, it follows that products of ANT-OAR would not be human organisms.

6.2.6 Induced Pluripotent Stem Cells (iPS cells)

These are obtained by manipulating somatic cells, essentially reprogramming them – in a way, de-differentiating them – back to a pluripotent state. A pluripotent cell is a relatively undifferentiated cell able to be coaxed into developing into other more specialized cells, almost any other cell of the body. IPS cells are obtained by taking a somatic cell, and introducing certain genes into its nucleus by viral vectors (most recently, by adenoviruses), with the result that the epigenetic state of the cell is changed from a specialized cell (say, a fibroblast cell) to a pluripotent cell. It seems that human iPS cells would have the properties that make embryonic stem cells desirable, without the ethical problems, and without many of the practical difficulties, of obtaining human embryonic stem cells. The grounds for classifying a pluripotent cell as a part of the human organism, rather than as a whole embryo itself, are similar to the grounds for classifying a skin cell as by nature a part [the skin cell is by its constitution oriented to being a part of a whole organism, though when isolated from an organism it is no longer actually a part – as a heart in an ice chest is not actually a part of anyone at that time, though by its constitution it is apt only to be a part] rather than a whole: the developmental trajectory is distinct from that of a zygote or embryo, and the genetic–epigenetic state of the cell is different from that of a zygote – its epigenetic state programs it to cooperate with other cells in the functioning of a larger organism of which it would be a part, rather than to develop itself to a more complex organism.

6.3 More and Less Severe Genetic Defects

From the preceding, it follows that it is possible to produce a cell or a group of cells that is human, in the sense that it is from humans and has the genetic sequence characteristic of humans in its nucleus, but that from the beginning lacks the genetic–epigenetic state that provides an active disposition to develop itself to the mature stage of a human being.

Of course, not every genetic or epigenetic defect means that those possessing it are not human organisms. Human embryos with Huntington's disease, Cystic Fibrosis, various types of Trisomy, and countless other genetic defects obviously have a genetic–epigenetic program orienting them toward developing themselves to the stage where they possess a brain suitable for being the substrate of conceptual thought, deliberate choice, and reflexive self-awareness. This is so even if the particular genetic defect an individual possesses also means that he will die before he matures very far. For example, the cause of most cases of anencephaly is environmental. But suppose (for the sake of argument) that some cases result from genetic defects. Still, these individuals are whole (though immature) human beings: it is part of their genetic–epigenetic program to develop whole brains, but they also possess a genetic defect in a different part of their genetic–epigenetic program (preventing the folding over of the neural tube in fetal development) that impedes the actualization of their intrinsic potential. Analogously, one's hand is designed to grasp objects, but it can also have a defect (for example, a broken bone) that prevents it from doing that. An organism can have a genetic–epigenetic program orienting it toward a certain mature stage, but also have a defect in another part of its genetic–epigenetic state that prevents it from fully actualizing his or her intrinsic potentialities.

This raises a further question. Would it be possible by ANT-OAR, or by a modification of parthenogenesis, to generate an individual that (a) was a distinct and whole organism, (b) was genetically human, (c) had a disposition to develop a brain, but (d) had a disposition only to develop the type of brain that is not suitable to be the substrate of conceptual thought – in other words, a brain but not of the right sort? The answer is that if such an individual *could* be produced, this would *not* be a human individual, a rational animal. This entity would not be an individual with the radical capacity for conceptual thought and deliberate choice – and so would not be a human being. It seems to me that to produce such an entity is *theoretically* possible. However, as long as an individual with the human genome has a brain, or the capacity to develop a brain, one could never verify with certainty that such an individual lacked the right sort of brain or the capacity to develop the right sort of brain (one suitable to be the substrate for conceptual thought and self-awareness). So, while I think this is theoretically possible, it is not practically possible: one could never be sure one had produced such an individual.

6.4 The First Four Days of Development

One objection to the ANT-OAR proposal – and by implication, to other proposals to obtain pluripotent cells – is that this procedure would in fact produce a disabled embryo rather than a non-embryo because in its early stages it was an embryo. The objectors admit that the expression of the Cdx-2 gene and the delay of the nanog gene's expression are crucial for the organization of the embryo. However, in the normal case of fertilization (or even in the "normal" case of cloning), the Cdx-2 gene is not expressed until the 16- or 32-cell stage, on approximately day 4. But that means (so it has been objected) that the failure in organization will not occur until approximately day 4. Thus, it appears that the organizational step that is prevented is not at the beginning (or possible beginning) of the embryo, but further down the line, sometime after the product of ANT-OAR has come into being. So, the objection is that the product of ANT-OAR appears to be a developing human organism between day 1 and day 4, and the deletion of the Cdx2 gene only causes this embryo to die at approximately day 4. Adrian Walker has argued that the process in question is accurately described as a "genetically programmed delayed structural collapse." [Walker (2005, p. 672). W. Malcolm Byrnes expresses this objection in the form of rhetorical questions: "The question here is this: Is not the embryo prior to the stage at which Cdx-2 expression is needed just like a so-called normal embryo, except for the silent 'defect' lurking in the genome? And, is not this situation precisely like that of a person with a genetic predisposition for Huntington's disease who lives a symptom-free life until the age of forty?" Byrnes (2005, p. 275]

First, to clarify the issue, it should be noted that the type of events that occur in the product of ANT-OAR that might appear to be internally coordinated also occur in masses that are generally admitted not to be unitary organisms. The first few rounds of cell division, with an enveloping zona pellucida (inherited from the oocyte) also occur in masses of cells that are clearly not, at any point in their proliferation, unitary organisms – for example, teratomas and CHMs. The manner of growth (including cleavage and compaction of the cells) is explained by the properties of the individual cells without the existence at any point of a higher level entity, that is, a whole organism.

Second, the decisive difficulty with this objection is that in an actual whole embryo the first holistic action, that is, the first action that cannot be adequately explained by the properties of the individual cells, is the differentiation of constituent cells into inner cell mass and trophectoderm. Yet, although the actual differentiation may not occur until day 4, the epigenetic alignments begin taking place from fertilization onward. The epigenetic alignments are the modifications of the various genes so that they will be expressed in the order and at the time that is needed for the first functional differentiation in the cells to occur. So, although the differentiation may not occur until day 4, this basic operation actually begins on day 1, since the epigenetic modifications preparing the way for this differentiation begin to occur at that time (Condic 2008). But if the Cdx-2 gene is deleted or silenced, and if the nanog gene is overexpressed, this basic operation does *not* begin on day 1, since the epigenetic factors that are actually the beginning of this operation cannot begin.

6.5 Evidence of Absence of Human Nature

A second objection that has been raised against ANT-OAR – but perhaps could be raised against the other proposals (parthenotes and iPS cells) – is that the proposal confuses the absence of evidence for the existence of a developing human organism with evidence of the absence of a human organism. As Adrian Walker expresses this objection: "While the presence of the appropriate genetic structure is a reliable sign of the presence of an organism of the corresponding sort, the absence of this structure is not necessarily a sign that the corresponding organism is absent, too." Walker (2005, p. 672)

This raises a legitimate question: there must be positive evidence of the absence of a human organism, not merely the absence of evidence. But it is important to see that the evidence that the product of ANT-OAR is not an organism is not simply that it fails to operate as an organism (and the same point applies to some parthenotes, iPS cells, etc.). That, of course, is *part* of the evidence, but it is true that an entity may fail to manifest its distinctive nature, either because of external impediments to the exercise of its powers, or – more significantly here – because of internal defects that are not so severe as to cause it to cease to be. But the evidence in the products of ANT-OAR, some parthenotes, and so on, in each case is more than the absence of organismal behavior. In addition to the absence of coordinated behavior along the trajectory toward the appropriate mature stage, it is known that the ingredients and the arrangement of these ingredients is such as to preclude from the beginning the development along the human trajectory. From the beginning, the materials lack what is necessary for the formation, that is, for the coming to be, of an animal-organism. For example, the deletion or silencing of the Cdx2 gene, plus the overexpression of the nanog gene, means that the entities produced are unable, from their beginning, to perform the first internally coordinated operation of the putative organism. So, this is not absence of evidence, but evidence of absence.

6.6 Mechanism?

A third objection is that the argument for the ANT-OAR proposal subtly presupposes a mechanistic view of material organisms – and perhaps the same might be said about the arguments for the other proposals. This objection has been expressed several times in slightly different manners. Adrian Walker claimed that: "[D]espite their [namely proponents of ANT-OAR] protestations to the contrary, they are mechanists who believe that living organisms are actually just complex machines: assemblages of parts whose unity is entirely a result of their completed assembly, and is in no sense 'more than the sum of its parts.'" (Walker 2005). This is because, argues Walker, proponents of ANT-OAR suppose that preventing the coordination of its parts into a whole will prevent the existence of the organism as a whole (what Walker calls "the coordinated all-at-onceness of its parts"). But this thought, says Walker, ignores the nonmechanist truth that the organism itself is the *cause* of the coordination of its parts. Only by confusing "the coordinated all-at-onceness" of an organism with the organism itself, the organism as a whole, says Walker, could proponents of ANT-OAR hold that suppression or expression of certain genes would be sufficient to prevent the coming to be of an organism.

David Schindler also advances this objection. He quotes a Joint Statement in favor of ANT-OAR (Production of Pluripotent Stem Cells by Oocyte Assisted Reprogramming: Joint Statement, http://www.eppc.org/publications/pubID.2374/pub_detail.asp) that was signed by various ethicists (myself included) and scientists when it says that, "the nature of each cell depends on its epigenetic state." Schindler then comments as follows:

> This claim, undefended in the statement, is what legitimates the proposal in the signatories' eyes. How? If the identity of a cell depends on its epigenetic state, and the totipotent zygote is a unicellular entity, then, so OAR's proponents reason, timely and strategic epigenetic modifications should suffice to prevent such a zygote from coming into existence in the first place (Schindler 2005, p. 371).

Later, Schindler characterizes this position as a mechanistic error. Rather, Schindler argues, epigenetics is not the first cause, but is only an expression of an already existing organism. According to Schindler, the proponents of OAR fall into mechanism by confusing a phenotype based on epigenetics with the substantial identity of the organism. [Describing the supposed basic error of ANT-OAR's proponents, Schindler sums up this objection as follows: "If, in fact, substantial identity is essentially a matter of epigenetics, then the absence or presence of an organism is simply a matter of reshuffling the epigenetic pieces – a classic mechanist maneuver. In a word, the assertion that OAR enables us to create pluripotent stem cells without creating an embryo is certainly true only if the mechanistic philosophy mediating this claim is certainly true, which it is not" (Schindler 2005 372).]

I certainly agree that mechanism is false. Mechanism is the denial of the existence of complex substances, the idea that what appear to be complex substances are actually only aggregates of substances. It is the idea that there are no wholes that are substantially more than the sums of their parts. As a machine is actually just an aggregate of substances, since the nature of each constituent is independent of its relation to other constituents in this group, so what appears to be a unitary substance composed of parts – an organism, for example – is (according to mechanism) really just a group of lower level entities arranged in a certain manner.

In contrast, nonmechanism holds that there are complex substances and that a whole organism, for example, has properties and behaviors not just resultant from the interaction among its parts. To take simple examples to illustrate the non-mechanist thesis: when a baseball shatters a window, it is actually just the atoms making up the baseball that shatter the window – the baseball in fact possesses no causal powers in addition to the causal powers of the atoms which together constitute what we call the baseball. So the baseball is only an aggregate of lower level substances, not itself a unitary substance. However, a genuine multicellular organism *does* have causal powers in addition to those of its constituents – for example, it grows, reproduces, maintains homeostatic conditions, and perhaps senses, performs voluntary movements, and so on. Walker and Schindler charge proponents of ANT-OAR with mechanism on the grounds that proponents of ANT-OAR hold that the modifications of the parts of constituents before they are joined can prevent the coming to be of a substance in the union of those constituents (and perhaps he would apply this criticism to proponents of other techniques as well).

But this charge involves a basic confusion about the nature of mechanism and about the proper role of material causes in any nonmechanistic understanding of material beings, including organisms. The nonmechanist recognizes formal and final causes as well as material and efficient causes (though this language may not always be used). In living beings the formal cause and final cause are the same – for example, a mature instance of an oak tree is the formal cause of the materials in the oak tree, but it is also the goal of its operations, the end to which it tends. Still, the material cause, that out of which a substance comes to be, is crucial in the generation of an organism – just as it also is crucial in the ceasing to be, that is, the death, of an organism. Thus, the proponents of ANT-OAR (and similar techniques) are not proposing, nor assuming, a mechanist view of organisms when they insist that an organism cannot come to be from materials that are not apt for the organization or coordination involved in a living material being of a specific kind.

It is an old saying that one cannot make a silk purse out of sow's ear – and that is a statement strictly about material causality. Even these days one could reverse that proposition – one cannot make a sow's ear out of a silk purse. (Even though, apparently, one *can* make a sow out of a sow's ear.) To take an example from the artificial realm: one needs a certain apt material in order to make a knife; one cannot make a knife out of clay or out of sand – the sand particles would not stick together sufficiently to have anything anyone would call a knife. Similarly, if the scientific claims made above about the Cdx-2 and the nanog genes are correct, then one cannot obtain an organism from the joining of a somatic cell with a silenced Cdx2 gene to an enucleated oocyte with an overexpressed nanog gene – the proliferated cells may adhere to each other, but they will from the beginning lack the capacity to function together for the first unified act of the organism as a whole. This position is not mechanism: it merely recognizes the importance of material causes.

In fact, the aptness of the material causes for the human form is the central issue also when questions about the death of a human being are considered. Whether brain death is the correct criterion of death or the cardiopulmonary criterion is the right criterion – in either case, what occurs in death is nothing other than the material constituents becoming so modified that they cannot maintain the unity or sustain the substantial form of a human organism. Thus, in both the coming to be and the ceasing to be of a human being, the right sort of materials must be present – and that fact provides a basis for a criterion for whether an unqualified coming to be or ceasing to be has occurred.

6.7 Stem Cells and Tetraploid Complementation

A fourth objection denies that stem cells, including iPS cells, are ontologically or morally different from human embryos – only, instead of concluding that both are human organisms, proponents of this objection conclude that neither are. I argued above that a human embryo is a human organism because it has the intrinsic potential to develop itself to the mature stage of a human being. But it has been objected that it has now become clear, through the process called tetraploid complementation, that stem cells, including iPS cells, also have the potential to develop to mature human organisms if they are provided the right environment, or provided an essential organ – namely, the trophoblast, from a tetraploid "embryo." In tetraploid complementation (a process used routinely with mice for genetic research), first, a tetraploid embryo is produced (for example, by fusing the cells in a two 2-cell embryo). The product is called a tetraploid "embryo," although its status is unclear (that is, it may not be a genuine embryo). [The status of any particular tetraploid "embryo" is ambiguous. When not complemented by embryonic or induced stem cells, tetraploid "embryos" often proliferate only trophoblastic tissue and in such cases clearly are not embryos. However, some (without the aid of stem cells) do develop, though more slowly, to mid-gestation, with all organs present, and in rare cases have even developed to term. See Eakin and Behringer (2003), Kaufman and Webb (1990), James and West (1994). So, some products of tetraploid complementation are genuine embryos, and for such a tetraploid embryo, the insertion of stem cells into it would be the insertion of stem cells into an ongoing embryonic development, rather than the generation of a new embryo. (I am indebted to Maureen Condic for alerting me to this point.) However, since this type of case apparently occurs infrequently, and those who have raised the objection considered above presuppose that tetraploid "embryos" are not whole organisms, I will focus mainly on the type of tetraploid complementation that involves a tetraploid "embryo" that is not a whole organism.] Next, a stem cell (either an embryonic stem cell or an iPS cell) is joined to this tetraploid "embryo," and (if the process succeeds) the resulting entity then develops as an embryo. In most cases of successful tetraploid complementation, the lineage of the cells in the mature fetus is entirely from the stem cells, while the placental and other "extraembryonic" tissues' cell lineage is entirely derived from the tetraploid cells. The objection is that this fact shows that the stem cells simply become the mature embryo and fetus and that the tetraploid "embryo" merely provides either a suitable environment or a needed organ (in effect, a transplant).

Thus, Gerard Magill and William Neaves argued as follows:

> A reprogrammed human cell is not fundamentally different from a nuclear-transfer or natural fertilization zygote in its ability to become a fetus. The zygote makes its own placenta, while the reprogrammed skin cell must be provided with one, but the placenta never becomes part of the embryo itself. Both the reprogrammed skin cells and the cells of the blastocyst's inner cell mass solely form the respective embryos. That is, reprogrammed skin cells have the same developmental potential as do the cells of the inner cell mass of the blastocyst formed by a zygote (Magill and Neaves 2009, p. 28).

This alleged equivalence is taken to be a *reductio ad absurdum* of the view that human embryos are whole (though immature) human organisms (Sagan and Singer 2007; Magill and Neaves 2009).

The first problem with this argument is that its central premise is inaccurate or misleading. The idea is that with tetraploid complementation the stem cells give rise to the fetus while the tetraploid cells give rise only to placenta. But, while in the mature fetus the lineage of all of the cells of the permanent body parts are from the embryonic or iPS cells, nevertheless, at earlier stages of gestation, both the tetraploid cells and the diploid cells are distributed throughout both regions. If the tetraploid–diploid (4n:2n) embryo dies at mid-gestation, many tetraploid cells are found within the permanent part of the embryo and diploid cells are found in the placenta or the placenta's precursor. Evidently, during several days after the complementation, the cells of the tetraploid "embryo" and the embryonic or iPS cells are distributed throughout all parts of the embryo but, perhaps partly because they divide more slowly, eventually the tetraploid cells migrate to the placental tissues (Eakin and Behringer 2003; Kaufman and Webb 1990). Hence, the evidence indicates that, in an embryo produced by tetraploid complementation, throughout its development the cells from the tetraploid "embryo" and the embryonic or iPS cells are intercommunicating and functioning as parts of a whole, self-developing organism.

Moreover, the placenta is clearly an organ of the embryo, albeit transient. The developing organism requires information provided by both inner cell mass cells and trophectoderm cells (the precursors to placenta). In prenatal life, the placenta functions as a vital organ of the embryo, sharing a common blood circulation with the rest of the developing embryonic body. (In contrast, the uterus of the mother is clearly part of a distinct organism that does not interfere with the internal functions of the embryo, but merely provides a supportive environment.) Hence, the iPS cells and the tetraploid cells together constitute an entire, integrated organism. At no point do iPS cells function as a whole organism, the way an early embryo behaves.

Another serious flaw in this argument can be seen by first considering a similar argument of which this is a descendent. Ronald Bailey and Peter Singer earlier argued that embryos are morally equivalent to somatic cells (such as skin cells) because somatic cells can produce mature human beings by way of cloning (somatic cell nuclear transfer, SCNT). In SCNT, the nucleus of a somatic cell is inserted into an enucleated oocyte and they are caused to fuse by an electrical stimulus, and the result (if all goes as planned) is a cloned embryo. Bailey and Singer argued that since somatic cells are converted into embryos, and these grow into mature members of the relevant species, human embryos are the ontological and moral equivalent of ordinary human somatic cells (Bailey 2001; Singer 2010).

However, when the nucleus of a somatic cell is joined to an enucleated oocyte, the enucleated oocyte does not act merely as an environment for the somatic cell that survives this change. Rather, when joined, both the somatic cell and the enucleated oocyte cease to be and their constituents enter into the makeup of a new entity. This is clear from the fact that the entities existing before the change (somatic cell and enucleated oocyte) are by nature parts of larger organisms, whereas what exists after the change is an entirely new entity, suited to act as a whole rather than as a part. In contrast, when an embryo develops, it survives and actively develops itself to its own internally directed maturation. Thus, human somatic cells have a fundamentally different status (ontological and moral) than human embryos.

The situation is similar with tetraploid complementation (provided the particular tetraploid "embryo" used is not already a genuine, though impaired, embryo). The stem cell (whether embryonic or induced) and the tetraploid "embryo" are causes (or co-causes) that generate a substantial change resulting in the coming to be of a new organism, that is, an embryo. If the tetraploid "embryo" were a mere environment, then the stem cells would persist and simply grow into, or proliferate, an embryo (and later a fetus). But what exists before the change (a disparate group of stem cells) has a different nature from what exists after the change (an entity that now has a radically distinct, and unified developmental trajectory). None of the stem cells, nor the stem cells as a group, was a whole organism prior to this fusion. Before their fusion, the stem cells were simply a disparate group of cells. After the fusion, the cells (both the stem cells and those derived from the tetraploid "embryo") function together to contribute to the maintenance and complex development of a new multicellular organism. Thus, the tetraploid "embryo" is not a mere environment enabling the stem cells to develop their latent potentialities. [Gerard Magill and William Neaves reply to this argument in Magill and Neaves (2009). They reply that it is true of both a zygote and an iPS cell in tetraploid complementation and that it interacts with a supportive environment: "In all cases, whether involving a zygote or an iPS cell, interaction with a supportive environment is required to express the developmental program residing in the genome." However, their reply mistakenly supposes that a cell's, or organism's, developmental potential resides only in its genome. But, while the genome is certainly important to developmental potential, the epigenetic state of that genome and cytoplasmic factors in the cell (or cells) are also essential to constituting what type of entity a cell (or group of cells) is. In the multicellular embryo, the three-dimensional arrangement of its cells is also important. The suppressed premise of their argument is that any two entities possessing the same genome possess the same integral active potency. But this is mistaken: it locates the integral potency of a cell or organism in a mere part (i.e., the genome) and ignores other crucial factors that help determine the active potential of that cell or organism. From the fact that a zygote and a stem cell are similar in some respects – the presence of either, together with other factors, can precede in time a developing embryo and then fetus – it does not follow that the zygote and stem cell are of the same nature. It was shown above, and elsewhere, that the zygote and the stem cell (whether embryonic or obtained from reprogramming) *combine* and *together* have a developmental pathway as a unit that is fundamentally different from the behaviors of either the stem cells or the tetraploid embryo prior to their junction. The behavior of this new organism (which includes both the stem cells and the cells of the tetraploid "embryo") is not merely the sum of the behavior of those cells, but involves an internally coordinated growth of a stable body including both types of cells.].

Still, it might be replied that this argument begs the question. For, two entities (say, what exists before tetraploid complementation and what exists after it) can behave in very different ways *if they are in different environments*. What exists before the tetraploid complementation (the stem cells) and what exists after may be of the same nature (and in fact the same being or beings) and yet behave very differently because they are operating in different environments. However, the evidence that a fundamentally new developmental trajectory has been initiated is not just that the (stem) cells behave radically differently when joined with the tetraploid "embryo," but that after their junction they *function together* with a direction of growth over and above that of the single cells, functioning *as parts of a new whole* – the actions of the organism as a whole embryo are clearly different from the sum of the actions of the disparate cells. As with naturally generated embryos, in these embryos there is intercommunication between the cells, orchestrating differential gene expression and patterned differentiated growth – formation of a body plan, gastrulation, and beginnings of organogenesis. This is evidence that what exists after the complementation (a whole composed both of the stem cells and the cells of the tetraploid "embryo") behaves in a fundamentally different way, and has a fundamentally different nature, than what existed before the complementation.

Again, it may be objected that the tetraploid "embryo" is like a transplanted organ. That is, it might be argued that the tetraploid "embryo" is a part that is needed for the survival and development of the stem cells (as a unit) on through to embryonic and fetal stages and that the addition of the tetraploid "embryo" does not change what it is inserted into. However, the group of entities it is joined to – the stem cells – either already constitute an embryo or do not. If they *do* constitute an embryo, this would mean that a group of pluripotent stem cells, *once detached from an embryo*, or produced by reprogramming, already constitute a whole organism, though lacking certain organs needed for survival. In effect, according to this alternative, detaching certain stem cells from an embryo would produce a whole, though debilitated and immature, human organism – somewhat like a cutting from a plant. And producing the same type of cell by reprogramming would also result in an embryo. [Agata Sagan and Peter Singer are mistaken when they claim that this idea leads to absurdities. They claim that to hold that such cells have moral status would lead to the moral obligation to split cells – repeatedly, or even indefinitely – from embryos in stem cell lines to protect their ability to develop. Sagan and Singer (2007, pp. 270–272). However, on the view they are criticizing such cells would have moral status – would be ontologically equivalent to human embryos – only after they were detached from the whole of which they previously were parts. Sagan and Singer consider something like this objection but reply that what was a part can be identical with what is now a whole – the cutting from a pelargonium, they say, is identical with the stalk and leaf or two it was when it was attached to that larger plant (Sagan and Singer 2007, pp. 272–274). But to hold that what was a part of an organism is now identical with a whole organism is to deny that the whole organism is something over and above a group of lower level entities. In effect, it is to deny that organisms are actually anything more than aggregates of particles (cells or perhaps molecules, atoms, etc.). This mechanistic position has several difficulties, one of which is that it is very difficult to accomodate our apparent awareness of self-identity through time. If what I am is an aggregate, then how is it the same I that deliberates about what to do tomorrow and that later carries out the choice I make? Dealing with such questions, a mechanist is pressed to identify the self with the experiences or memories, or with a subject of experiences, both of which would be other than the organism (which, after all, on this view, would have to be only a series of particles in motion, or a type of process, rather than a persisting entity). For more on this issue, see Oderberg (2007); Connell (1988), and Lee and George 2008b), Chapter 1. I do not think the alternative that the detached human stem cells (or iPS cells) are actually human embryos is correct, but Sagan and Singer have failed to show that it leads to absurdities.] If this were true, then the objection that the iPS cells and embryos are ontologically equivalent would no longer be a successful *reductio ad absurdum* – rather, one would say that detaching stem cells at a very early stage (*embryonic* stem cells) or producing their epigenetic equivalent by reprogramming (as in iPS cells), would actually be much the same as detaching a totipotent cell from a four-cell embryo, a process that produces a new whole embryo. [If I understand him correctly, Hans Werner Denker suggests that this may be the case with iPS cells Denker (2009).] On this view what are called iPS cells (induced pluripotent stem cells) actually constitute a distinctive type of clones.

But this alternative does not seem to fit the facts. The epigenetic state of naturally occurring stem cells must be significantly changed in order for such cells to form a unified, whole organism. But no such change occurs with their detachment. Also, the induced pluripotent cells have an epigenetic state equivalent to that of natural stem cells, and both have material constitutions disposing them to function only as parts of wholes. In contrast, a dramatic change does occur with the junction of the tetraploid "embryo" to the embryonic or iPS cells, after which they function together as a single organism.

If, on the other hand, it is argued that the stem cells before being joined to the tetraploid "embryo" do *not* constitute an embryo, and also that the tetraploid complementation does not generate an embryo, then one must explain at what point, and how, several disparate cells become unified with no apparent cause. The joining of the stem cells to the tetraploid "embryo" causes a radically new developmental pathway that: (a) is distinct from that of either of the entities joined, (b) belongs to these cells as a unit, and (c) from this point on has the same pattern as other embryos, namely, a regularly recurring, complex, and apparently interiorly coordinated process of growth resulting in the mature stage of an organism of the relevant species. This regularity and internally orchestrated unity in the early development of the embryo remains unexplained on the view that an organic unity arises only gradually at some point during gestation.

The fact that in tetraploid complementation the mature embryo's cell lineage is completely derived from the stem cells proves nothing concerning what it is, or whether it is identical with an earlier group of cells or not. It certainly does not show that the embryo or fetus is the same entity or group of entities as those stem cells. An organism's clone has its entire cell lineage from its one parent, but it is, of course, a distinct organism.

The evidence, then, shows that what exists before the tetraploid complementation (the stem cells, and the tetraploid "embryo") are different in nature from what exists after the change. Hence, tetraploid complementation is not the provision of a mere environment for the development of latent potentialities in stem cells, nor merely the transplantation of a missing vital organ. Rather, it is the generation of a new entity, a new organism. This new organism, the new embryo, is not ontologically or morally equivalent to a stem cell (embryonic or induced) but has the ontological and moral status of a human organism at the early stages of its development.

In short, a human organism comes to be with fertilization, and probably also in the production of a partial hydatidiform mole (though surviving only a short time), and perhaps also with some induced parthenotes (some induced parthenogenetic processes involving alteration of gene expression). However, CHMs, some parthenotes, products of altered nuclear transfer (with ooctye-assisted reproduction), and induced pluripotent stem cells are not human organisms, and therefore one could (if it is scientifically feasible) ethically derive pluripotent stem cells from them.

6.8 When There Is Doubt

Finally, I would like to address briefly the question of what should be done if one cannot in a given case be certain that an entity is or is not a human embryo. What moral judgment should be made about the production and manipulation of such an entity? What must one do in case there is reasonable doubt? The answer is that if there is reasonable doubt about the nature of what one is considering manipulating or killing, then one ought not to do it. There are two reasons for this. First, justice demands that we not kill what we have reasonable doubt may be a human being: each of us hopes and morally demands that others be reasonably certain that we are dead before we are dismembered or destroyed, and the golden rule requires that we treat others as we would have them treat us.

Second, the case here is quite distinct from choosing to perform an action which one knows has some risk – in that case one might accept that risk as a side effect, depending on how great the risk is, how great the burden of omitting the action, one's special responsibilities if any to those who might be adversely affected by one's action, and so on. But here we are not choosing to perform action A, foreseeing that it might cause B (the risk). Instead, we are considering choosing to kill a particular entity, with the knowledge that killing this entity might be killing a human being. So, if we choose to destroy this entity, then we are consenting to kill what this entity might be. In other words, if one chooses to kill what might be a human person, then one consents to killing a human person. So, if there is reasonable doubt about whether a biological entity is or is not a human person, one morally must treat it as a human person.

References

Austriaco N (2005) Are teratomas embryos or non-embryos? A criterion for oocyte-assisted reprogramming. Natl Cathol Bioeth Q 5:697–706

Bailey R (2001) Are stem cells babies? Only if every other human cell is too. Reason online, July 11, 2001, http://reason.com/archives/2001/07/11/are-stem-cells-babies. Accessed 23 Jan 2010

Brevini TAL, Gandolfi F (2008) Parthenotes as a source of embryonic stem cells. Cell Prolif 41:20–30
Byrnes WM (2005) Why human 'altered nuclear transfer' is unethical, a holistic systems view. Natl Cathol Bioeth Q 5:271–279
Condic M (2008) Alternative sources of pluripotent stem cells: altered nuclear transfer. Cell Prolif 41:7–19
Connell RJ (1988) Substance and modern science. Houston, TX, Center for Thomistic Studies
Denker HW (2009) Induced pluripotent stem cells: how to deal with the developmental potential. Ethics Biosci Life 4:34–37
Eakin GS, Behringer RR (2003) Tetraploid development in the mouse. Dev Dyn 228:751–766
Grisez G, Lee P (2011) Total brain death: a reply to Alan Shewmon. *Bioethics* (in press)
Huarte J, Suarez A (2004) On the status of parthenotes. Natl Cathol Bioeth Q 4:755–770
James RM, West JD (1994) A chimaeric animal model for confined placental mosaicism. Hum Genet 93:603–604
Kaufman MH, Webb S (1990) Postimplantation development of tetraploid mouse embryos produced by electrofusion. Development 110:1121–1132
Lee P (2004) The pro-life argument from substantial identity: a defense. Bioethics 18:249–263
Lee P, George RP (2008a) The nature and basis of human dignity. Ratio Juris 21:173–103
Lee P, George RP (2008b) Body-self dualism in contemporary ethics and politics. New York, Cambridge University Press, pp 133–140
Magill G, Neaves W (2009) Ontological and ethical implications of direct nuclear reprogramming. Kennedy Inst Ethics J 19:23–32
Oderberg D (2000) Applied ethics. Oxford, Blackwell, pp 31–41
Oderberg D (2007) Real essentialism. Routledge, New York
Paffoni A et al (2007) In vitro development of human oocytes after parthenogenetic activation or intracytoplasmic sperm injection. Fertil Steril 87:77–82
Sagan A, Singer P (2007) The moral status of stem cells. Metaphilosophy 38:264–284
Schindler D (2005) A response to the joint statement, 'production of pluripotent stem cells by oocyte assisted reprogramming. Communio 32(Summer):369–380
Singer P (2010) The revolutionary ethics of embryo research. Project syndicate, December 2006, http://www.utilitarian.net/singer/by/200512–.htm. Accessed 23 Jan 2010
Suarez A, Lang MD, Huarte J (2007) DIANA anomalies: criteria for generating human pluripotent stem cells without embryos. Natl Cathol Bioeth Q 7:315–335
Walker A (2005) Altered nuclear transfer: a philosophical critique, Communio 31: pp 675–676

Chapter 7
On the Status of Human Embryos and Cellular Entities Produced Through ANT: Are They Persons?

Pablo Requena Meana

Abstract The term "embryo" is used in biology to refer to the early developmental phase of an organism. In philosophy, "person" is employed to designate the specific dignity of those beings that possess self-consciousness, freedom and moral responsibility. Presently, there is a great debate on the possibility of granting the category of persons to human embryos. Such a step would make them subject of rights and thus prevent their destruction. In this chapter, we will show the complex nature of these debates in the field of bioethics given that the problem of the epistemology status of bioethics has yet to be resolved. We will also analyze the category "person" and the various meanings it has acquired in philosophy over the centuries. In the concluding part, we will present the Catholic doctrine on this theme and on some recent proposals for obtaining embryonic stem cells (like in ANT). In order to do so, we will rely primarily on the Instruction *Dignitas personae*, the latest document issued by the Church on the topic.

Keywords ANT • Catholic Church • Embryo • Person

7.1 Introduction

In order to understand the title of this chapter a preliminary comment is necessary. In the last decade, a lot has been written on alternative means for obtaining stem cells, either for research or for treatment, without having to destroy embryos. In the decade prior, the bibliography devoted to the use of human embryos for investigation is almost inexhaustible. While the majority of articles dwell on the scientific aspects, some however make reference to the philosophical and ethical character of the process. In this article, we propose to tackle the issue of the ontological status of

P. Requena Meana
Pontifical University of the Holy Cross, Rome, Italy
e-mail: requena@pusc.it

A. Suarez and J. Huarte (eds.), *Is this Cell a Human Being?*,
DOI 10.1007/978-3-642-20772-3_7, © Springer-Verlag Berlin Heidelberg 2011

the embryo and of the cellular "entities" derived from altered nuclear transfer (ANT), using the rest of the contributions in this book as our reference point.

If one bears in mind the ethical problems related to the various proposals around the personal status of the embryo, the enormous difficulty in realizing our objective can be appreciated. Some decades ago, this difficulty was inexistent since there was then no possibility of manipulating embryos in the laboratory as can be done now. The question as to whether the embryo (or the fetus) was a person remained at a more theoretical level. It was usually limited to the discussions on abortion. Some people, even though they were willing to accept the personhood of the embryo, tried to justify abortion by invoking a supposed conflict of rights. With the arrival of artificial reproductive techniques, the issue has become more relevant since it would be difficult to deny the embryo the rights of a person, especially the right to life, if it is indeed regarded as a person. The changes that would need to be introduced in the health system and in scientific practice would lead to such a revolutionary restructuring that some have postulated that the situation already supposes a convincing argument for reducing the negation of the personhood of the embryo to the absurd (Wall and Brown 2006). The authors are right in pointing out that many people in the scientific field do not regard the embryo as a person and that many current practices would need to be radically changed but this tells us nothing about the personhood (or not) of the embryo from a purely logical point of view. On another note, it is important to point out (although we cannot fully develop the argument here) that it is not necessary to prove the personhood of an individual before a demand can be made for some rights to be accorded to him or her. Perhaps, a very clear example is the Universal Declaration of Human Rights.

This chapter is divided into three parts. In the first part, the issue to be treated is introduced from the perspective of the epistemology status of bioethics. Next, the central topic, which is the philosophical concept of person, will be tackled as well as its possible application to the human embryo. A classical concept, like that proposed by Boethius, which is based on the categories of substance and (rational) nature would consider human embryos as persons. A modern concept, for example that of Locke which is founded on self-consciousness, would have no place for the embryo, which at most would be considered as a "potential person." At the same time, we will briefly mention the issue of "delayed hominization" supported by Thomas Aquinas and other medieval authors and which has been used by some to deny the personhood of the embryo based on a classical concept of person. Finally, we will dedicate a section to a presentation of the directives given by the recent Magisterium of the Catholic Church on the status of the embryo and the other products derived from ANT.

7.2 The Status of the Embryo in Bioethics

One of the most prominent characteristics of bioethics, right from its beginnings, has been its multi-disciplinary nature. The resolution of some ethical issues which have arisen in the area of biotechnology (especially its application to human beings) needed an approach that covered a wide range of perspectives. Not only were biological and medical data needed, but also juridical, ethical and anthropological reflections were necessary in order to better understand the problems and give better-founded solutions. This characteristic that has accompanied bioethics from its beginnings has also turned out to be one of the greatest challenges that the new discipline faces. What methodology should it adopt, among the ones used in the different fields which constitute it? What importance should the scientific conclusions have on the ethical solutions offered? There still remain many unanswered questions and the epistemology status of bioethics still remains unsatisfactorily resolved. This however has not prevented some authors from formulating articulated proposals which have enjoyed a wide diffusion (Beauchamp and Childress 2008; Jonsen et al 2006; Sgreccia 2007).

With regards to our study, we distinguish four different but inter-related levels in which the premises, arguments, and conclusions which arise in discussions about the human being can be classified. At the biological level, answers are sought to the question about the biological nature of the embryo. The metaphysical nature of the embryo is considered at the philosophical (or ontological) level, while the best way of treating the human embryo is usually discussed at the ethical level. Finally, legal questions regarding the rights and juridical protection that should be accorded to the embryo are answered at the juridical level.

All these levels are inter-related. The ontological status of the embryo, for example, is considered beginning with the results obtained at the biological level. Nevertheless, it would be misleading to arrive at conclusions about the former from the data obtained from the latter without further analysis. It is necessary to go beyond the biological level if philosophically founded conclusions are to be drawn. Confusing the different aspects considered at the various levels has led to methodological errors and to poorly justified conclusions. Two examples can serve to illustrate this point. Bole (1990) writes that it is impossible empirically to justify (in a non-religious way) that the zygote has the same principle of psychological integration (the soul) as the adult clearly does. Of course it cannot be empirically demonstrated by biology. Neither can the soul of the adult be shown at the biological level. One has to go over to the philosophical level in which questions regarding the integrating principle of a living human organism are asked and in which the existence of an immaterial element called the soul is postulated. It is only at this level, and without any recourse to religious reasons, that one can question whether this immaterial principle can also be applied to the embryo. The second example is taken from the already-quoted article of Wall and Brown (2006). The authors write that affirming that an embryo is a person on the basis of the genetic potential of the human zygote is a poorly justified way of using genetics. It also implies an ignorance of the extraordinarily inefficient and wasteful nature of human reproduction. Certainly the authors are correct in saying that the personhood of the embryo cannot be based only on genetic data. It is necessary to find out, from a philosophical perspective, the implications of this genetic potential in order to ascertain whether it has any importance to the question being discussed.

7.3 The Person as a Metaphysical Concept

7.3.1 From the Biological to the Ontological Level

Having highlighted the dangers inherent in confusing the different levels in bioethical discussions, we will now consider the ontological status of the embryo. This consideration must, by necessity, begin from the data available at the biological level (what the Ancients called physics) in order to arrive at the metaphysical level which transcends the former. If a given philosophical investigation is based on an inadequate representation of the topic under study (which in this case is given by biology), a perfectly logical argument can be developed without it having any connection with the real world. It would remain, at most, as an interesting mental construction. Thus scientific investigation is important and, in not a few cases, the bioethical argument on a particular issue rests fundamentally on the data provided by science. An example can be seen in the case of the neurological criterion for the diagnosis of death. Many writings tackle philosophical questions related to death, the integration of the entire person and his possible interaction with the outside world, etc. However, these discussions are of no use in determining whether the criterion of brain death is adequate from an ethical point of view if one did not have a previous explanation of what happens at the physio–pathological level. Neither can they be of help in knowing if one cannot inductively arrive at the establishment of a criterion that may give (medical) certainty, as is the case with the cardiocirculatory criterion for the determination of death.

Carrasco de Paula (1987), while speaking on the ontological status of the embryo, puts forward two examples in order to show how an erroneous biological concept leads to philosophical conclusions that are baseless. On the one hand lies the theory known as "preformation" which supposes a perfectly differentiated organism from the very moment of fertilization and which leads the philosopher to conclude that the rational soul is infused in the very moment of conception. On the other hand, another theory considers a succession of life principles in the developing human embryo: at first, it has a vegetative soul, followed by a sensitive (or animal soul), and finally an intellective or human soul. In the process each successive soul could carry out the functions ascribed to the previous souls. On the basis of this latter theory, the philosopher would then be in favor of delayed ensoulment. The problem in both cases does not lie in the falsity of the conclusions (one of which must be true) but rather in the fact that they have been reached through an insufficient methodology.

After showing the importance of the biological data, we will use the biological status of embryos and other entities derived from ANT found in the chapters this book that directly deal with the theme as the foundation for our analysis. We will make particular reference to the conclusions of Condic (Chap. 3) who writes that the "zygote formed by fusion of sperm and egg is a human organism, i.e. a human being at the earliest stage of development." For now, we will focus only on the first part of the phrase: the zygote or unicellular embryo (and thus the developing embryo) is an organism of the human species (*Homo sapiens*). It is a scientific conclusion that few experts today deny.

7.3.2 Philosophical Characterization of the Human Embryo

Returning once more to the phrase of Condic just cited, we have to point out that even though there is a consensus on its first part, the same can hardly be said for the second part. She writes that the zygote is a human being from the very first stage of development. It is an assertion that goes over to the metaphysical plane since "human being" is not, strictly speaking, a term that is proper of biology in the same way that "human organism" is. It rather belongs to the field of philosophy. Some authors have distinguished between three concepts – "an individual of the *homo sapiens* species," "human being" and "persons" – affirming that they are not equivalent terms (Álvarez 2005). In the opinion of these authors, there are persons that are not humans and there are individuals of the *homo sapiens* who lack some characteristics that are proper of humans. Those who fall into this last group should be considered neither as human beings nor as persons. The author just quoted includes embryos in the last group. He does not however derive any ethical conclusions as necessary consequences of this distinction at the ontological level, quite unlike what is common among other proposals advanced in bioethics. He is of the opinion that a moral and legal protection, equivalent to that required for persons, can be demanded for those individuals of the human species that do not enter into the category of persons.

The following metaphysical question has been put forward: "what type of being is the human embryo?" Depending on the answer to this question, it may or may not be possible to respond to the thesis of Álvarez. In order not to unduly prolong this presentation, we take for granted, as something commonly agreed upon, the affirmation that although not all persons belong to the category of human beings (God and the angels for example), that which makes an individual of the *homo sapiens* species to be a "human being" (as understood by Álvarez) is what also characterizes the individual as a person. Thus the question that has to be answered is whether we can apply to the embryo what is specific to a person. But then, what marks an individual as a person? The topic of our study revolves around this question. As can easily be perceived, the problem lies in the fact that there is no single answer. In the course of the history of philosophy, various responses have been given to the question.

For the sake of simplicity, we can distinguish two main currents of thought in the way person is conceived. There is a classical view, whose paradigm can be found in the definition of Boethius: "individua substantia rationalis naturae" (individual substance of a rational nature). This definition which has its origins in the theological debate between Boethius on the one hand and Eutyches and Nestorius on the other at the beginning of the sixth century would latter be widely used in medieval philosophy. A second view is the modern concept of person of which Locke can be considered a representative author. He considers self-consciousness, which makes an individual to remember the past and question himself with regards to the future, as the basis for the identity of the subject. It is a concept of person defined in functional terms in relation to legal imputability.

The classical view of person is thus based on the metaphysical concepts of individual substance and rational nature. The application of "individual substance" to the embryo has been objected to on the grounds that it should not be considered as an individual, at least in the period in which it is possible for twinning to occur. Nevertheless, this objection does not appear well-founded since a single individual can be conceived which afterwards divides into two or more individual substances due to internal or external factors. In zoology, there are numerous examples of asexual reproduction which take place in this fashion. Even though human reproduction is sexual, this does not prevent the possibility of a certain "asexual reproduction" in the early phases of embryo development through a division of the morula into two genetically identical embryos that develop as two different organisms. Part of the reason behind the objection is the confusion between individuality and indivisibility, due, to a great extent, to the metaphysics of the monad of Leibniz. A response to the objection is that what characterizes an individual is his actual individuality and not his potential indivisibility (Ide 2007).

Of greater interest for the consideration of the human embryo as a person is the second category used by Boethius in his definition: rational nature. We believe that this concept (with nature understood in the classical sense of "essence as principle of operations") is what permits a human organism in which all the characteristics proper to humans are not fully present to be considered a person. The fact of having this nature (rational, human) and not another, is what enables such an individual to be included within the group of persons. This criterion which is present in all the individuals of the human species removes the distinction suggested by Álvarez between an individual of the species *homo sapiens* and a human being. All human organisms are human beings since they all have a rational nature, even though it may not be manifested externally, or even when it is impossible that it will ever be manifested. This way of thinking has greatly contributed to the moral progress of the western societies and has led for example to the abolition of grave racial discriminations.

The concept of person as self-consciousness lies at the base of some of the present proposals in bioethics, for instance in the functionalist-actualist proposal of Engelhardt (1996) who defines the person on the basis of characteristics like self-consciousness, autonomy and rationality. In this way of considering a person, it is easy to realize that it is impossible for the embryo to be included within the concept of person. However, as Rodríguez Luño (2008) rightly points out, this theory reduces substance to function *in actu*. This debatable metaphysical posture is then used to determine the moral status of the embryo, which, as can be easily deduced, is left in a totally unprotected situation. The underlying problem of such moral systems is that the will of the strong is imposed on the weak, since the former establish the criteria which individuals of the human species have to fulfill before they can enjoy the moral and juridical status of person.

7.3.3 The "Human in Potency," Aristotle and Thomas Aquinas

It is important to dedicate some lines to evaluate the thesis that argues that the embryo is not a person but that it can eventually become one. Some authors, basing themselves on Aristotle and medieval scholasticism, write that the embryo is a "human in potency" (a potential human being) and not actually one. One of the first persons to support this thesis in bioethical discourse was Ford (1991) who saw it as compatible with the guidelines of the Warnock Report on the individuality of the embryo which was thought to take place only after the first 14 days.

We will make use of the works of Berti (1993, 2007) and Rodríguez Luño (2008) in order to respond to Ford's thesis. The concepts of matter and form and in particular, the necessity of a certain organization in the matter that would enable it to receive the form, lead Aristotle to postulate a type of progressive reception of the form by the product of conception. The male semen contributed the principle that transmitted the specific form but the material principle provided by the woman could not receive a sensitive or intellective soul from the very beginning even though they were present in potency. This epigenetic theory adopted in the theological field by some, among them Thomas Aquinas, led them to postulate that the human soul could only be infused by God to the fetus after 40 days for males and after 80 or 90 days for females. This doctrine forms the foundation for the denial of the personhood of the embryo by some contemporary theologians.

The above thesis can be answered from two different perspectives, one biological, and the other philosophical. The knowledge of the beginnings of human generation (and of animals in general) possessed by Aristotle and the medieval authors was quite limited. The mere consideration of their ignorance of the gametes and their role in human reproduction is enough to illustrate this. Neither is there any biological basis for their distinction between male and female fetuses. This leads us to believe that Aristotle, Thomas Aquinas and the medieval authors would not have written the same thing if they had possessed the same knowledge which we have today. This can be demonstrated if we take the philosophical presuppositions behind their arguments into account.

Actually, for Aristotle, the human embryo already contains the intellective soul in "first act" although it is still incapable of putting its own faculties into "second act" as it has only the vegetative faculties in "second act" (*De anima*, II, 3: 412 a 23–29). As the embryo develops, so do the organic components that will enable it to carry out its operations (the sensitive ones first and the properly human ones afterwards). Therefore, it cannot be affirmed that for Aristotle the early embryo can only be considered as a simple plant, or that it only possesses a vegetative soul, which successively is substituted by a sensitive soul and much later, by an intellective one. From the beginning, the embryo has a rational soul (that which is proper to man) which potentially contains the vegetative and sensitive souls, much in the same way that a polygon contains a triangle and a square within. For Aristotle on the other hand, the male semen contains the formal principle or the development plan, the program which Berti identifies with the DNA at the biological level, even though the latter does not exhaust the formal principle of the male gamete. In the opinion of the Aquinas, the male semen cannot yet be considered a "man in potency" (in the sense of possessing the intellective soul in first act) since it needs to be placed in another being and to be transformed by another principle (*Metaphysics* IX, 7: 1049a 14–16).

For this line of thinking to be understood, it is necessary to distinguish between the concepts of potency and possibility. In the case of potency, the final result is due to the activity carried out by the subject itself and it is normally arrived at if no obstacle is met. Possibility on the other hand refers only to the non-impossibility of becoming one thing or another, but the outcome depends on factors external to the subject. In our opinion, it can be presumed that for Aristotle, the phrase "human in potency" does not mean, as its present linguistic use may suggest, that an entity is not yet human. Rather, it means that the rational faculties are in potency, a situation that is different from the rational soul which is already in (first) act.

With regards to the view of Thomas Aquinas, we rely on the excellent work of Pangallo (2007). In order to understand why the Angelic Doctor defended the delayed hominization of the embryo, it is necessary to consider the theological discussions that were going on in his time. In particular, he wished to refute the "translationist" thesis according to which the rational soul is transmitted from father to child through semen. Aquinas wanted to defend the direct creation of each human soul by God, infused into a matter that is capable of receiving the soul (with an adequate "dispositio corporis" as he termed it). Using the knowledge of biology available in his time, he thought that this corporal disposition did not exist at the beginning of the generative process but was only acquired at a latter moment. Pangallo does not hesitate in asserting that, combining the current knowledge we have of biology with the metaphysics of Aquinas (especially his distinction between *actus essendi* and *existentia*), it is possible to sustain the immediate hominization of the embryo, and consequently, its personhood. At the existential level, a gradual material formation of the body, which takes place in different phases, can be observed. However, at the level of being, it can be maintained that the rational soul (which includes the vegetative and the sensitive souls) is infused at the beginning of this process and acts as the *actus essendi* of the new being right from the start. Eberl (2006) reaches the same conclusion when he writes that "the presence of brain's epigenetic primordium is thus sufficient for a human zygote to have active potentialities for a rational soul's proper operations, since the zygote's ordered natural development will result in an actually thinking rational human being."

7.3.4 Can All the Spontaneously Lost Embryos Be Regarded as Persons?

One of the objections raised against the personhood of the embryo which seems, at least intuitively, to have more weight, is the large number of embryos that are normally lost in the reproductive process of animals, humans included (Wall and Brown 2006). This objection had already been formulated in a provocative manner many years ago by the theologian Karl Rahner: "How can it be thought that 50% of human beings – having immortal souls and destined for eternity – never get beyond a priori this first stage of human existence?" (Rahner 1967). We shall set aside the issue regarding the actual number of embryos that perish, a datum whose verification poses undeniable empirical difficulties. Although we do not know with absolute certainty the number of embryos that are lost (Huarte and Suárez 2004), nevertheless we know that they form a high percentage and this makes Rahner's question to be of utmost relevance. However from the logical point of view, the objection lacks adequate basis since the fact that a phenomenon occurs frequently tells us nothing about the subject or the object involved. The infant mortality at the beginning of the twentieth century was around 10% but this high percentage reveals nothing about the personal nature of neonates.

All things considered, I believe that Rahner's concern can be resolved by taking Aquinas' aforementioned concept of "dispositio corporis" into consideration. Many scientific studies have shown that an elevated number of the products of fertilization contain chromosomal anomalies, or at least some genetic aberrations which are incompatible with life (Huarte and Suárez 2004). The majority of these cellular entities do not survive beyond the first stages of cellular division. Thus we can say that many of them never get to become human embryos in the strict sense of the term although they may divide in a similar way as the latter and may even acquire the form proper to a blastocyst. This means that not all the products of fertilization end up as embryos. From a philosophical perspective, and taking into account the matter/form distinction mentioned above, we can affirm that the biological products (which may be called the matter) do not attain a sufficient make-up or format (an adequate "dispositio corporis") to be informed by a rational soul (the form that is proper to the nature to which this living being belongs to). This might explain why many of the products of fertilization that perish are not human embryos and therefore are not persons.

Affirming that a number of the products of fertilization will not develop into embryos may appear to weaken the protection due to the early human organism. However, this is not the case. From the ethical point of view, there seems to be no problem with the utilization (including the destruction in the investigative process) of these embryonic-like products of human fertilization. There are however two difficulties that need to be overcome. On the one hand lies the ethical issue bound to the techniques of in vitro fertilization, the only way of obtaining embryos in the early stages of development outside the mother's body. On the other hand remains the technical difficulty of discovering a valid criterion for distinguishing between an embryo and what is an embryo-like entity. This latter problem may be overcome with more advances in embryology but the difficulties involved should not be underestimated.

7.3.5 Altered Nuclear Transfer: Between Biology and Metaphysics

To this point, we have been mainly concerned with human embryos. Now, we shall say a few words on the biological entities obtained from ANT, a technique that has been proposed as a valid option for procuring embryonic stem cells without the need for destroying human embryos.

This technique was presented to the President's Council on Bioethics in 2004. It consists in the activation of an ooplasma (an ovum which has had its nucleus removed) by the insertion of the altered nucleus of a somatic cell. In principle, the silencing and/or activation of certain genes of the inserted altered nucleus prevent the formation of a human embryo as a result of the ovular activation (Hurlbut 2005). The "biological artifact" (following the terminology used by the White Paper issued by the President's Council) resulting from this insertion would merely be a group of pluripotent stem cells, without the internal structure or organization proper to the human embryo. This "artifact" would therefore lack any kind of teleological tendency to develop as a human specimen. This has led Hurlbut to conclude that ANT is an ethically valid alternative to the use of embryos in the attainment of pluripotent stem cells. Oocyte assisted reprogramming (OAR) is a particular type of ANT in which some of the genes that are expressed in pluripotent cells (but not in totipotent ones) are activated.

When considering the proposals made by defenders of ANT, and specifically the arguments that the "product" of ANT is not a human embryo, an important terminological problem arises in addition to the purely biological one. The roots of this terminological problem are found in the philosophy of nature and more specifically in the philosophy of biology. In one of the most cited articles in this debate, ANT is described as having produced pluripotent stem cells from deficient blastocyts of mice (Meissner and Jaenisch 2006). From the very beginning of the contemporary debate on the nature and direction of embryonic development, no one doubted that the formation of a blastocyst was a stage in embryonic development, in both humans as well as many animal species. In the present case, however, this term has come to mean a model or type of cellular structure (rather than a definite developmental stage), which may or may not be an embryo (Hurlbut et al. 2006). These authors write that: "The term 'embryo' is generally defined as the human organism from fertilization to the end of the eighth week. 'Blastocyst' is a structural description designating a fluid-filled spherical form, and is generally used in embryology to indicate the stage between morulation and gastrulation. Blastocyst-like structures, however, are common in a range of tissues during organizational development, and therefore this term can be used without implying the presence of a living organism." This shift in meaning could simply be considered as a better definition of the terms used within a particular science (embryology). However, our impression is that it is related to a deeper question, one which reappears in much of the literature generated by the ANT debate. It is bound to the fundamental issue that we have been considering in this chapter of the relationship between biology and metaphysics.

Some of the writings on ANT make transitions from the biological to the philosophical realm that appear, to say the least, somewhat abrupt and forced. In the case of ANT, the difficulty involved in making properly metaphysical judgments is that the subject matter in question is not easily observable "by all" or by the majority of people. This problem is only exacerbated when one considers that the biological phenomena under discussion involve minor changes at the molecular level (the activation and deactivation of one or a few genes, the silencing of certain proteins, etc.). This makes it very difficult to determine whether or not the so-called "biological artifacts" (the matter in philosophical terms) are capable of being informed by a spiritual soul. We are not here denying the possibility of drawing valid ontological conclusions from the study of descriptive data provided by molecular biology. Nevertheless, it seems clear that caution must be exercised in this matter, especially if one considers the experience provided by the history of science. As is well known, many scientific postulates have been modified owing to contemporary advances in molecular biology. We can think, for example, of some of the initially unquestioned findings of Ramon and Cajal in the field of neuronal regeneration which have subsequently been revised in the light of more recent stem cell research. Philosophers and scientists, therefore, should recognize that this is a field in which it is extremely difficult to define or delineate the precise subject matter under consideration.

7.4 The Ethical Criterion of the Magisterium of the Catholic Church with Reference to Human Embryos and ANT

In the final section of this chapter, we will discuss the issue of the ethical status of the embryo through a presentation of the related Catholic doctrine. Although this teaching appears clear and articulated, it is often presented as having no role to play in public debate. Rather it is treated simply as a one valid religious perspective among others.

7.4.1 Catholic Magisterium and Science

When the Magisterium has had to give instructions on ethical issues that are intimately related to the problem of identifying a thing's biological nature, it has made reference, whenever possible, to the current state of science on the particular topic. In some cases, its moral judgment is subject to possible scientific advances and clarifications. Two examples may help to illustrate this point. The Instruction *Donum vitae* (1987) speaks about the subject matter closely related to our study: the scientific, ontological and moral characterization of the embryo and the beginning of human life. In that Instruction, we find a particularly illuminating passage: "the Church's Magisterium does not intervene on the basis of a particular competence in the area of the experimental sciences; but *having taken account of the data of research and technology*, it intends to put forward, by virtue of its evangelical mission and apostolic duty, the moral teaching corresponding to the dignity of the person and to his or her integral vocation. It intends to do so by expounding the criteria of moral judgment as regards the applications of scientific research and technology, especially in relation to human life and its beginnings" (italics added).

The second example comes from the Church's input concerning the use of neurological criteria for determining death. In August of 2000, Pope John Paul II addressed the XVIII International Congress of the Transplantation Society. In his address, he explained that when it comes to establishing solid scientific guidelines for determining death, "the fundamental criterion must be the defense and promotion of the integral good of the human person, in keeping with that unique dignity which is ours by virtue of our humanity." He then went on to note that "it is a well-known fact that for some time certain scientific approaches to ascertaining death have shifted the emphasis from the traditional cardio-respiratory signs to the so-called "*neurological*" *criterion*. Specifically, this consists in establishing, according to clearly determined parameters *commonly held by the international scientific community*, the complete and irreversible cessation of all brain activity (...) With regard to the parameters used today for ascertaining death – whether the "encephalic" signs or the more traditional cardio-respiratory signs – *the Church does not make technical decisions*. She limits herself to the Gospel duty of comparing the data offered by medical science with the Christian understanding of the unity of the person." John Paul II also stated that the criteria for neurological determination of death do not appear to be inconsistent with Christian anthropology and therefore may be adequate for achieving the degree of moral certainty required by Magisterial teaching for acting in a morally – upright way *vis-a-vis* organ transplantation. The wording of the Pope's argument clearly leaves open the possibility of revising the moral evaluation of "brain death" criteria should medical science reach conclusions different from those that were assumed to be true at an earlier time. In that case, one would need to reconsider the problem under the perspective of those advances.

7.4.2 Catholic Teaching on Human Embryo

The Church's opposition, right from its early years of existence, to whatever form of voluntary abortion is well known. This has been so even when the precise facts regarding the biological nature of the embryo in its first stages of development (as we presently understand it) were unknown (Sardi 1975). The condemnation of voluntary abortion is based on the commandment "do not kill" which is applied to all the range of human existence.

With reference to the first phases of embryonic development, which is the focus of this chapter, we shall now present some texts of the Magisterium of the Catholic Church which address the ontological and ethical specifications of the embryo. Our aim is not an exhaustive study of the texts but only a limited selection of the relevant ones.

We find one of the first ethical considerations in a speech of Pius XII directed to a group of Italian doctors in 1944: "As long as a man is not guilty, his life is untouchable, and therefore any act directly tending to destroy it is illicit, whether such destruction is intended as an end in itself or only as a means to an end, whether it is a question of life *in the embryonic stage* or in a stage of full development or already in its final stages" (italics added). For its part, the Second Vatican Council affirms in one of its documents that "life must be safeguarded with extreme care *from conception*; abortion and infanticide are abominable crimes." (*Gaudium et spes* 1965, italics added).

The Instruction *De abortu procurato* (1974) of the Congregation for the Doctrine of the Faith explains that "the first right of the human person is his life," adding that there is no justification for any discrimination based on the different periods of life. The Instruction further remarks that "in reality, respect for human life is called for from the time that the process of generation begins. From the time that the ovum is fertilized, a life is begun which is neither that of the father nor of the mother, it is rather the life of a new human being with his own growth. It would never be made human if it were not human already." This assertion was made based on the findings of modern genetics. At the same time, it explains that "it is not up to biological sciences to make a definitive judgment on questions which are properly philosophical and moral such as the moment when a human person is constituted or the legitimacy of abortion." One can clearly see the distinction which the Church's Magisterium makes in this text between the two planes of the problem: science should shed light on issues related to biology without attempting to answer (from its own area of specialization) the philosophical question about the personhood of the embryo or the embryo nor the ethical one regarding the legalization of abortion.

Some years later the already-mentioned Instruction *Donum vitae* (1987) on the respect owed to nascent life and the dignity of procreation appeared. In its first part entitled "Respect for human embryos," the Congregation for the Doctrine of the Faith offers an ethical guideline: "The human being must be respected – as a person – from the very first instant of his existence." Afterwards, it explains that "Certainly no experimental datum can be in itself sufficient to bring us to the recognition of a spiritual soul; nevertheless, the conclusions of science regarding the human embryo provide a valuable indication for discerning by the use of reason a personal presence at the moment of this first appearance of a human life: how could a human individual not be a human person?" The rhetorical question brings to mind the proposal of Álvarez that has been already analyzed and which does not follow the position defended by the instruction since it distinguishes between "individual of the *homo sapiens* species," "human being" and "person." The Instruction, for its part, taking into account the current philosophical debate over the concept of person, remarks a little later on that "The Magisterium has not expressly committed itself to an affirmation of a philosophical nature, but it constantly reaffirms the moral condemnation of any kind of procured abortion."

This teaching was further ratified by the encyclical of John Paul II, *Evangelium vitae* (1995). After quoting the Instruction *Donum vitae*, he writes that "what is at stake is so important that, from the standpoint of moral obligation, the mere probability that a human person is involved would suffice to justify an absolutely clear prohibition of any intervention aimed at killing a human embryo. Precisely for this reason, over and above all scientific debates and those philosophical affirmations to which the Magisterium has not expressly committed itself, the Church has always taught and continues to teach that the result of human procreation, from the first moment of its existence, must be guaranteed that unconditional respect which is morally due to the human being in his or her totality and unity as body and spirit." Independent of any particular philosophical conception of personhood, this teaching of the Church continues to serve as an ethical guide for the respect owed to every human being.

The most recent magisterial document on this issue is the Instruction *Dignitas personae* (2008), published by the Congregation for the Doctrine of the Faith. This document, which aims at updating the teachings contained in *Donum vitae*, continues the same line of argumentation as the previous instructions. In the text of *Dignitas personae*, one can appreciate, on the ontological level, a greater closeness to the thesis regarding the personhood of the embryo even though it is not explicitly defined. Actually number 5 of the document points out that "In fact, it presupposes a truth of an ontological character, as *Donum vitae* demonstrated from solid scientific evidence, regarding the continuity in development of a human being. If *Donum vitae*, in order to avoid a statement of an explicitly philosophical nature, did not define the embryo as a person, it nonetheless did indicate that there is an intrinsic connection between the ontological dimension and the specific value of every human life. Although the presence of the spiritual soul cannot be observed experimentally, the conclusions of science regarding the human embryo give "a valuable indication for discerning by the use of reason a personal presence at the moment of the first appearance of a human life: how could a human individual not be a human person?" Indeed, the reality of the human being for the entire span of life, both before and after birth, does not allow us to posit either a change in nature or a gradation in moral value, since it possesses *full anthropological and ethical status*. The human embryo has, therefore, from the very beginning, the dignity proper to a person" (italic from the original).

We can thus conclude this part of the write-up by saying that the Magisterium of the Catholic Church, using the data offered by modern sciences and bearing in mind the philosophical debate over the ontological status of the embryo as well as the different concepts of person which exist, offers an articulated ethical guideline: the embryo deserves the protection that is due to a human person.

7.4.3 Altered Nuclear Transfer in Dignitas Personae

The just-mentioned Instruction *Dignitas personae* is the only Church document that discusses ANT and other possible ways of obtaining stem cells. In this concluding part, we shall look at the guidelines which it offers for the moral evaluation of the technique.

Dignitas personae specifically mentions two techniques which involve the induction of parthenogenesis in human eggs: altered nuclear transfer (ANT) and

OAR. The instruction explains that "these proposals have been met with questions of both a scientific and an ethical nature regarding above all the ontological status of the 'product' obtained in this way." The document goes on to state a clear ethical principle which applies to all such techniques, "while these questions are not clarified, we should note the following statement in the encyclical Evangelium Vitae, 'what is at stake is so important that, from the standpoint of moral obligation, the mere probability that a human person is involved would suffice to justify an absolutely clear prohibition of any intervention aimed at killing a human embryo'." Faced with such a strongly worded statement, we might get the impression that there is little else to say or discuss on the matter as far as *Dignitas personae* is concerned. It remains necessary, however, to examine more deeply the scientific, ethical, and ontological questions raised by the above mentioned techniques in order to provide sound moral advice to scientists involved in this area of biomedical research.

Before continuing our discussion, it is important to recall that the techniques under consideration are the product of highly innovative scientific research undertaken to discover treatments for a large number of hitherto incurable diseases. Another important point is that they have been developed as alternative methods to those techniques which involve the use and destruction of human embryos – an action repeatedly condemned by the Church. While discussing advances in medical science more generally, the same Instruction states the fundamental moral criterion by which to judge medical progress: "These developments are certainly positive, and they deserve support, as used to overcome or correct pathologies and help to restore the normal development of generative processes. They are however negative and therefore cannot be adopted, when they involve the removal of human beings, are using means that are prejudicial to the dignity of the person, or taken for purposes contrary to the integral good of man."

The fact that *Dignitas personae* condemns violations of human dignity involved in certain medical developments should not suggest that there is a general feeling of suspicion in the Church towards scientific and medical innovation. In fact, it would be contrary to the Christian spirit if researchers did not diligently seek new methods or procedures to help combat disease. Therefore, in general, research proposals that are not contrary to the principles of justice are to be positively valued by Christians. Of course, the concrete decision to carry out a particular line of research needs to be evaluated on a case-by-case basis. This evaluation will take into account, among other factors, the risks of the proposed research, the likelihood of success, and the costs of implementing it. In this decision-making process, however, it is important not to confuse prudence with omission. The fact that a certain line of research could present some ethical difficulties does not mean that we should reject it without further in-depth study of the ethical questions it raises.

In the case of the specific techniques mentioned above (ANT and OAR), the central question is whether or not they violate the norms of justice. As a springboard for discussing this question in the light of *Dignitas personae*, I will briefly present William Hurlburt's discussion of ANT. In Chap. 10 of this book, he offers ten moral principles intended to delineate possible ethical boundaries for hybridization, cellular gene cloning projects, and stem cell research. The first principle specifically excludes the production of normal human embryos in such research, that is, of any entity possessing an internal self-capacity to develop into the more mature stages of human life. The second principle calls for caution whenever there is any ambiguity about the status of the body in question or the identity of the species. Within this framework, Hulburt argues that it is ethically acceptable to use the research technique ANT when striving to produce pluripotent stem cells. In the nearly five years that have elapsed since ANT was first proposed as an alternative method for the acquisition of pluripotent stem cells, several works have examined its ethical status (Austriaco 2006; Condic 2008; Hurlbut 2007; Berg and Condic 2008). For the purposes of this chapter, I will not repeat all the objections that have arisen concerning this technique. My remarks will be limited to the treatment of ANT by *Dignitas personae*.

A first point to note is that *Dignitas personae* recognizes that ANT is a technique that has been specifically developed in order to avoid the need for therapeutic cloning and the destruction of human embryos in stem cell research. However, the Instruction also situates its discussion of ANT within a larger section devoted to human cloning (and not in another section which addresses the use of stem cells). This particular place ANT occupies within *Dignitas personae* has important consequences regarding both ANT's characterization as a technique and its moral character. The inclusion of ANT within a larger section on human cloning seems to deny, or at least call into question, precisely what the defenders of ANT have argued from the beginning, namely that the technique does not produce a human embryo and is therefore not a form of human cloning (although the technical manipulation of the egg is almost identical to the usual techniques of cloning by nuclear transfer). As regards to its moral character, the placement of ANT in *Dignitas personae* alongside ethically problematic practices like reproductive and therapeutic cloning makes it hard for ANT to avoid a certain shadow of suspicion or "guilt by association" arising from its proximity to these practices. Of course, the mere location of ANT amidst certain unethical practices is not a definitive argument against the technique. In my opinion, however, it shows, at least in part, the attitude of *Dignitas personae*'s authors towards ANT.

In the few lines devoted by *Dignitas personae* to the subject of ANT, mention is made of the controversial questions that the latter poses for science and philosophy. The document states that while this technique has generated various scientific and ethical questions, the main problem it raises is a metaphysical one. This problem can be expressed in the following way: what "ontological" status should be attributed to the biological entity that results from ANT? There are only two possible answers to this question. The product of ANT is either an individual of the human species in its early stages (i.e., an embryo) or a live cellular structure which lacks essential characteristics proper to the human organism (something akin to human cancerous tissue). If the first option is true, then the technique cannot be justified for any reason since the destruction of embryos is always unlawful as a violation of the most elementary principle of justice, i.e., the right to life. The use of human tissue in medical research, on the other hand, does not usually raise any particular moral problems. We may therefore conclude that the problem posed by

ANT is primarily scientific in nature: one must determine whether or not the "biological artifact" which results from the technique is a human embryo.

Dignitas personae also mentions the existence of other ethical questions related to ANT. It does not, however, specify their nature. I think there are two other possible moral objections worth considering: the use of human ova in ANT and the possibility of scandal that accompanies the technique. Some authors have expressed doubts about the morality of using human reproductive material (sperm and ova) in research even in cases in which no embryos are formed. From this perspective, the whole "status" question is bypassed and the objection to ANT becomes an objection to its illicit use of human oocytes. On this question, the Instruction *Dignitas personae* offers only little guidance. While it positively condemns ova cryopreservation, it does so in reference to in vitro fertilization and not to experimentation with eggs per se. The question this raises is whether it is morally acceptable to use germ cells for purposes other than procreation. We think the answer to this question is yes. It should be noted that, by its own dynamics, oocyte activation initiates a series of mechanisms that are teleologically ordered towards the formation of a human embryo. However, their use in research or treatment should not depend on fostering this "natural" course. Having said that, there is no reason why a reproductive cell, or its use, should have a moral characterization distinct from other cells in the body. Its use (in ANT) could be considered analogous to transplanting tissue for gonadal hormone replacement therapy (to treat an ovarian deficiency), which is considered legitimate in principle, and which differs from the intended transplant of human eggs into a person for the sake of reproduction. Faggioni (1998) has aptly traced the history of these practices which began in the late nineteenth century and their moral evaluation. The main objections raised were related to the morality of transplants *ex vivo*. Other objections focused on the risks involved for the donors as well as the purpose of the intervention (for example, it was generally not considered acceptable to undergo such a transplant simply to recover one's sexual capacities). Additionally, Faggioni mentions another relevant ethical consideration, namely the possible effects of the procedure on the recipient's personality. It should, therefore, respect the ethical criteria to be observed by any ex vivo organ donation, so long as it avoided significant risks to the donor's health and any commercialization of human eggs of persons with limited financial resources (President's Council on Bioethics 2005).

The second objection, the danger of scandal, is mentioned four times in *Dignitas personae*, each time in relation to the problem of cooperation in evil. Nevertheless, the Instruction only mentions scandal in cases involving the use of illicit materials (as in the use of cell lines or materials derived from the destruction of human embryos or voluntary abortion). The scandal mentioned in the document therefore involves a different scenario than that presented by ANT. The possible danger of scandal in the case of ANT is not linked to the tainted origin of research material, but rather to the confusion it could engender in the minds of non-experts, which in this case, would be the great majority of persons. This confusion might arise from its technical similarity to cloning and in vitro fertilization (with which it shares several notable features). Within the perspective of *Dignitas personae*, therefore, neither the technique's use of human eggs nor the possibility of scandal seem to constitute definitive objections against the use of ANT. They should be taken into consideration, however, when giving a complete moral evaluation of ANT.

7.5 Conclusion

In the debates surrounding the morality of the use of embryos for biomedical research, it is common to find assertions and arguments which are located in four distinct planes – biological, metaphysical, ethical and juridical. Bearing in mind that each plane makes uses of a specific methodology, it is obvious that errors can arise in the arguments adduced due to the ease of passing from one plane to another without sufficient basis. With regards to the human embryo, there is a great unanimity on the part of experts that it can be considered an organism of the human species. The problems arise when one moves on to the philosophical plane and asks if the embryo can be considered a person. We have seen that the answer to this question depends on the philosophy that one adopts and that in the history of human thought, the understanding of "person" has changed from a concept based on substance to that based on function. In the former, it is easy to include the embryo in the category of persons. It is however more difficult, if not impossible, to do the same using the latter concept. All things considered, the philosophical discussions on the way of understanding what man is or his ontological characterization should not be used as an excuse for lessening the moral and juridical protection to some members of the human species. A reasonable doubt whether an individual is a person or not should be sufficient to justify the moral prohibition of its destruction. This has been the method adopted by the Magisterium of the Catholic Church, which without taking sides in the philosophical discussions, has however always defended the protection of every human being from the very first moment of his existence.

Acknowledgments I wish to express my appreciation to Profs. Rodríguez Luño and Jiménez Amaya for their comments and observations. This study was translated into English by Anthony Odoh and Craig Iffland. I also thank Kevin Belgrave for his revision of the manuscript.

References

Álvarez JC (2005) Ser humano-persona: planteamiento del problema. In: Masiá J (ed) Ser humano, persona y dignidad. Universidad Pontificia Comillas-Editorial Desclée De Brouwer, Bilbao, pp 17–41

Austriaco NPG (2006) The moral case for ANT-derived pluripotent stem cell lines. Natl Cathol Bioeth Q 6:517–37

Beauchamp TL, Childress JF (2008) Principles of biomedical ethics, 6th edn. Oxford University Press, Oxford

Berg TV, Condic ML (2008) Emerging biotechnologies, the defense of embryonic human life, and altered nuclear transfer. Linacre Q 75:268–290

Berti E (1993) Quando esiste l'uomo in potenza? La tesi di Aristotele. In: Biolo S (ed) Nascita e morte dell'uomo. Problemi filosofici della bioetica, Marietti, Genova

Berti E (2007) Individualità biologica e artificio. In: Gessa-Kurotscha V (ed) Saperi umani e consulenza filosofica. Meltemi Editori, Roma

Carrasco de Paula I (1987) Personalità dell'embrione e aborto. In: AAVV Persona, verità e morale. Città Nuova Editrice, Roma

Condic ML (2008) Alternative sources of pluripotent stem cells: altered nuclear transfer. Cell Prolif 41(Suppl 1):7–19

Eberl JT (2006) Thomistic principles and bioethics. Routledge, London

Engelhardt HT (1996) The foundations of bioethics, 2nd edn. Oxford University Press, Oxford

Faggioni M (1998) I trapianti di gonadi: storia e attualità. Med Morale 48:15–46

Ford NM (1991) When did I begin?: Conception of the human individual in history, philosophy and science. Cambridge University Press, Cambridge

Huarte J, Suárez A (2004) On the status of parthenotes: defining the developmental potentiality of a human embryo. Natl Cathol Bioeth Q 4:755–770

Hurlbut WB (2005) Altered nuclear transfer as a morally acceptable means for the procurement of human embryonic stem cells. Perspect Biol Med 48:211–28

Hurlbut WB (2007) Ethics and embryonic stem cell research: altered nuclear transfer as a way forward. BioDrugs 21:79–83

Hurlbut WB et al (2006) Seeking consensus: a clarification and defense of altered nuclear transfer. Hastings Cent Rep 36:42–50

John Paul II, Address to the 18th International Congress of the Transplantation Society (29.08.2000). http://www.vatican.va/holy_father/john_paul_ii/speeches/2000/jul-sep/documents/hf_jp-ii_spe_20000829_transplants_en.html. Cited Feb 14 2011

Jonsen AR et al (2006) Clinical ethics: a practical approach to ethical decisions in clinical medicine, 6th edn. McGraw Hill, New York

Meissner A, Jaenisch R (2006) Generation of nuclear transfer-derived pluripotent ES cells from cloned Cdx2-deficient blastocysts. Nature 439:212–215

Pangallo M (2007) The philosophy of Saint Thomas on the human embryo. In: Sgreccia E, Laffitte J (eds) The human embryo before implantation: scientific aspects and bioethical considerations. Libreria Editrice Vaticana, Vatican City

Rahner K (1967) Zum Problem der genetischen Manipulation. In: Rhaner K (ed) Schriften zur Theologie, Bd VIII, Einsiedeln

Rodríguez Luño A (2008) Scelti in Cristo per essere santi. III Morale speciale. Edusc, Roma

Sardi P (1975) L'aborto ieri e oggi. Paideia, Brescia

Sgreccia E (2007) Manuale di Bioetica, 4th edn. Vita e Pensiero, Milano

The President's Council on Bioethics (2005) Alternative sources of human pluripotent stem cells. A White Paper. Washington DC, pp 36–48

Wall LL, Brown D (2006) Regarding zygotes as persons: implications for public policy. Perspect Biol Med 49:602–610

Chapter 8
Does a Human Being Have a Right to Life? The Debate on Embryo Research in Germany as a Case Study

Manfred Spieker

Abstract This chapter discusses whether a human being has a right to life and analyzes the political, philosophical, and social conflict between two undeniable goods: "freedom of research" and "embryonic protection". The frame of reference for the discussion is the contemporary debate over embryonic experimentation in Germany. I identify the relevant historical and legal particularities of the German debate, but also attempt to highlight how the issue has been framed by different parties to that debate. The result is a study in how the particularities of a country's social and political history can help to frame the debate over the moral status of human embryos and the limits of science in interesting and complex ways.

Keywords Biopolitics • Embryo • Human being • Property rights • Right to life

8.1 Introduction

In summer 2000, the debate about bioethics in Germany was opened by the Feuilleton of the German newspaper FAZ. On the 31st of May 2001, the German Bundestag had its first debate about stem cell research, followed by its second great discussion on 30th of January 2002, which moved the nation, such that it passed the stem cell law on the 25th of April 2002. On the 11th of April 2008, the Bundestag extended the number of embryonic stem cells that could legally be imported from those derived by the year 2002 to those derived up to 2007. The matters of that debate were changing: whereas at the beginning the issues were stem cell research and the corresponding therapeutic visions of physicians and biologists, later the main matters were pre-implantation genetic diagnosis (PGD) and cloning; on the

M. Spieker
University of Osnabrück, Osnabrück, Germany
e-mail: mspieker@uni-osnabrueck.de

A. Suarez and J. Huarte (eds.), *Is this Cell a Human Being?*,
DOI 10.1007/978-3-642-20772-3_8,

20th of February 2003, the Bundestag took a resolution to prohibit the latter, independent from its reproductive or therapeutic target.

The crucial questions in all these fields of discussions have always been the same. What is the moral status of the embryo? When does human existence begin and when does the protective obligation by the state start? The answer given by the German embryo protection law is unequivocal. It is also the answer corresponding to the observations of natural science and the jurisdiction of the German Federal Constitutional Court. The human life begins at the moment when there is the fusion of ovum and sperm. The fact that at that moment also protective obligations by the state have to begin is no result of religious belief but the condition of legitimacy of a secular state. That is how the legislation regarded the case in the embryo protection law and how the Federal Constitutional Court in its judgements according to abortion – although the explanations by the Court concerning the realization of that protective obligation resulted in irresolvable contradictions (Spieker 2008, Spieker 1993).

In the scientific debate about bioethics where physicians, bio-chemists, lawyers, philosophers, theologians, sociologists, literary scientists, Enquete commissions, and ethic councils are participating in, the positions are more manifold but they deal with exactly the same crucial questions: Which is the moral status of the embryo? When does human being begin and when does the protective obligation by the state begin? The controversial as well as manifold answers to these crucial questions can be subdivided into two camps, independent from the scientific discipline of the participating people: the camp of advocates of consumptive embryo research, pre-implantation diagnosis and therapeutic cloning, and the camp of their adversaries.

8.2 Positions of Advocates of a Consumptive Embryo Research

Among the protagonists of consumptive embryo research there is the neuropathologist Oliver Brüstle who refers to his experiments with mice cells which have opened perspectives “to get the crucial problems of transplantation medicine under control in long-term view”. That is why “ethically and medically it is inexcusable” not to continue this research. If few blastocysts are sufficient “to develop new strategies to treat patients having diseases which today still are incurable”, you should weigh between the duty of the physician to heal on the one hand and antiquated ethical conceptions of values on the other hand. He says that we are not allowed to block the access of patients to new therapies. The ability of infinite multiplying and pluripotentiality make embryonic stem cells “an inexhaustible donor source for transplantation medicine; neurocytes for patients having Parkinson, cardiac muscle for victims of myocardial infarct, insulinogenic cells for diabetics and hematogenic cells for patients having leukemia are only some of the visions connected to the new technology.” (Brüstle 2000; Diedrich 2001; Bartram 2001) That is why the experiments with embryonic stem cells involve considerable economic interests and corresponding patent applications are an "ethical obligation", according to Brüstle (Brüstle and Wiestler 2001). To admit that above all they include the death of the embryo is astonishingly difficult for Brüstle.

This position is supported by the biologist Hubert Markl. For him, the fertilized ovum "still does not mean that this is a human being" (Markl 2001a, 2002, p. 45). The fusion of ovum and sperm does not constitute more than a genetic program. To see the human being himself just in possessing a set of human genes, Markl says, is a mere "biologism" (Markl 2001b). The idea of man is a "culture-related conception of attribution" (Markl 2002, p. 45). To be entirely a human being man has "to transgress his limits". He is "talented with human dignity especially by the fact" that in his decisions he does not submit to natural facts like the moment of fusion of cell nuclei (Markl 2002, p. 49). That is why embryonic research whose affirmation results from the duty of the humans to "rule their own destinies" is not an "expression of the hubris of progress made by blinded scientists but an indispensable part of our human nature and human dignity" (Markl 2002, pp. 54–55).

On the 3rd of May, 2001 the Deutsche Forschungsgemeinschaft DFG (German Research Association) had revised its former refusal of research with embryonic stem cells and recommended the admittance of research with imported embryonic stem cell lines, an active participation of German scientists in their production and the change of the embryo protection law which is necessary for that admittance. The Association is convinced that these recommendations "on the one hand correspond to our comprehension of constitution and perception of law, on the other hand also to a vision of man which does justice to the scientific research as well as to the legitimate interests of ill people" (DFG 2001).

Constitutional arguments for that change of policy by the DFG are above all delivered by Rüdiger Wolfrum. He says that it is not allowed to close by law the way to possibly promising therapies. Because of that fact and because of the constitutionally granted freedom of research the development of therapeutic measures must not be restricted if no higher-ranking legally protected interest is endangered. The questions of whether the life of an embryo is a higher-ranking legally protected interest or not and whether we may strive for a maximum of health at the expense of the life of a third person are avoided by distinguishing between embryos which are created in vitro and embryos growing in the mother's womb, thereby denying the status of being human to the embryo in vitro. "The argumentative equation of embryos outside and inside the mother's womb supersedes the importance of the mother for the development to a human and overemphasizes the importance of the genetic disposition. This does not do justice to the phenomenon of becoming a human being and of individuality. The existing differences between embryos inside and embryos outside the mother's womb justify a constitutional consideration which includes those differences." (Wolfrum 2001a, b, 2002; Koch 2001; Schäuble 2001).

Another distinction which has the purpose of balancing legally protected interests and to allow research with embryonic stem cells is the distinction between the "existing or growing human being" on the one hand and the "human life" on the other hand, used specially from professors of Lutheran Theology. Whereas it is conceded that the existence and growth of human beings entails that they are worthy of protection, this is denied to embryonic human life. It is true that the latter "in principal could have become a human being". But without a uterus there is no growing human but only a surplus embryo. "By this, the decisive reason to hinder the research with those embryos is inapplicable." That is why the research with surplus embryos is "morally justifiable and compatible with the Christian conception of human life." (Fischer 2001, 2002).

Also among representatives of the pharmaceutical industry there are pleas for a consumptive embryo research, which are extremely bound by special interests. Daniel Vasella, the President of the Novartis Group, does not shrink from positing his own anthropology to prove his position: "Who may claim to know exactly at what moment human life begins and whether this is a gradual or an abrupt event?" In his opinion and in contrast to the ignorance he declared immediately beforehand, he then asserts it is clear that becoming a human is not abrupt but a gradual event and that there is no dignity of an integral, developed human being for an embryo, and so he has "no principal scruples" against consumptive embryonic stem cell research. He says that we have to weigh the existing risks against the chances to heal and that we must not hinder progress because of any personal conviction (Vasella 2002). Actually, the logic should result in the opposite conclusion. If I do not know the moment when human life begins I am obliged to protect also its earliest phase. The same conclusion must be made if I claim that becoming a human is a gradual event.

8.3 Positions of Adversaries of a Consumptive Embryo Research

Wolfgang Huber, Lutheran bishop of the region Berlin-Brandenburg-Schlesische Oberlausitz until 2009, objects to the aforementioned thesis regarding the distinction between embryos in vitro and embryos in the mother's womb as follows: "Our protective obligations towards the embryo are growing in the same measure as our possibilities to intervene are increasing." The fact that the embryo in the mother's womb before nidation is relatively unprotected, he says, is "no justification for the conclusion that the researcher can do with the embryo in the Petri dish whatever he likes to do". In the Petri dish the researcher disposes of protective possibilities that are not available in the mother's womb. This conviction also is decisive for the embryo protection law (Huber 2001). But the thesis of Huber can not only be proven by the fact that the technology of reproductive medicine facilitates the protection of the created embryo, it is also based on the logic of the social state of law whose protective obligations are growing according to the degree of dangers. It is a command of justice to give a special care to the embryo in vitro.

For all persons criticizing consumptive embryo research, human life does not start with nidation but with the fusion between ovum and sperm. "The natural finality of the fertilized human ovum is a given condition of law which is not submitted to a valuation at will. That is why the embryo is under the protection of the guarantee of human dignity." (Starck 2001) According to recent research this finality can be considered present at the moment the sperm penetrates the egg (Rager 2004; Condic 2008). "Once this has happened, both sperm and oocyte undergo such significant changes that neither seems to exist in its own right anymore" (George and Tollefsen 2008). Even if the German Grundgesetz admits interventions because of a law in the field of the right to live such that the protection of life experiences legal graduations, the guarantee of human dignity which does not admit any graduation prohibits any instrumentalization of the embryo. That is why the so-called surplus embryos must not be made available to serve the purposes of others. The human dignity commands to let them die. Accordingly, their "useless" death is no "senseless" death (Hillgruber 2002; Classen 2002).

Nevertheless, it has been argued that one cannot infer from the judgements of the Federal Constitutional Court that "the protection of the Art. 1 par. 1 GG begins only with nidation. However, what both senates said in 1975 and 1993 can be transferred to the judgment about the embryo from the moment of fertilization." (Benda 2001). So, the embryo in vitro is no less a holder of the fundamental right to live and of physical integrity than the embryo in the mother's womb. The embryo has the same guarantee of human dignity, which prohibits any instrumentalization for the purposes of third persons. And also the artificially created embryo who can no longer be implanted into the donor of the ovum, independent from the reasons, must be protected against lethal instrumentalization. "The proximity of death does not make the affected persons corpses, and the lack of chances to continue life does not include the allowance to start a life-destroying intervention... And so there never can exist an obligation of stem cell research for so-called surplus embryos" (Höfling 2001).

Already for Günter Dürig in his commentary to the German GG 1958 the prohibition regarding instrumentalization constituted the core of the guarantee of human dignity (Dürig 1958). That is what leads numerous lawyers, moral theologians, philosophers, and biologists to reject PGD too. As to the medical lawyer Adolf Laufs, in the diagnostic selection there is "a clear violation or instrumentalization of the human embryo", which injures his dignity as well as his right of physical integrity (Laufs 2000; Röger 2000; Beckmann 2001; Böckenförde 2003). So it contradicts to the Grundgesetz and the embryo protection law. For the moral theologian Eberhard Schockenhoff, the ethical principle "to accept every other person in his individual existence" does not endure any weakening, even if there are "location considerations in research policy or economic utility being opposed to that". But in PGD and therapeutical cloning the embryo is "instrumentalized in a manner which contradicts the acceptance of human dignity, the right to live, and of physical integrity". So, the duty to omit that instrumentalization has priority over the duty to contribute to the realization of "principally desirable targets like the fulfillment of a wish to have a child or the treatment of diseases which are not treatable until now". In the case of a conflict, the negative obligations to omit have priority over the positive obligations to act (Schockenhoff 2001a, b; Spaemann 1999; Höffe 2001; Spieker 2011).

Finally, for the philosopher Robert Spaemann the absolute respect "independence from the context" of the person is the reason why we are never allowed to use others only as a means for our purposes. This absoluteness forms the frontier of our duty to help. "Rescuing as many lives as possible can justify the sacrifice of one's own life, but not the intended killing of an innocent" (Spaemann 1996). The status of a person depends on the fact "that this status is not given but that every person in virtue of his own right enters the group of persons". If human rights were "given" or "conceded", "those rights would not exist, because then it is a question of defining power to whom those rights are given and to whom they are not given" (Spaemann 2001).

With regard to the human embryo, also Jürgen Habermas states a prohibition against instrumentalizing. For him, PGD, embryonic stem cell research and interventions into the human genome are violating the "conditions of reciprocity of communicative understanding" which in the case of all interventions in genetic technology and medicine requires the possibility of assuming a consensus given by the affected embryo (Habermas 2001, pp. 90–93). For Habermas, human dignity is bound to the "symmetry of relations". "It is not a quality you can possess by nature like intelligence or blue eyes; it rather marks that inviolability which can only have a meaning in the interpersonal relations of reciprocal acceptance, in equal association of persons with each other." (Habermas 2001, p. 62). PGD, embryonic stem cell research and interventions into the human genome violate these interpersonal relations and provoke a "different self-understanding in generic-ethical aspect which can no longer be harmonized with the normative self-understanding of self-determined living and responsibly acting persons" (Habermas 2001, p. 76). But the consistency of his refusal of interventions using genetic engineering in the human embryo is weakened by Habermas himself when he distinguishes "between pre-personal human life" and person (Habermas 2001, pp. 62, 72). Accordingly, he sets a difference "between the dignity of human life and the human dignity granted by law to every person." By no means, "the genetically individualized creature in the mother's womb" is a person. It is true that it has an "integral value for the whole of an ethically composed form of life", but the "individuation of his life's history" is not be made before the "socialization" (Habermas 2001, pp. 64, 63, 96). With this socialization of the person, in the tradition of Hegel or Marx, where the substance dissolves in the relations, the birth becomes the decisive point for a human to become a human. Accordingly, Habermas calls it the "moment of entering into the social world...from that the talent to be a person is able to realize."

Regine Kollek and the political scientist Ingrid Schneider suppose the change of policy of the DFG to be a fight for patent rights and material interests which are concealed by their protagonists. It is true that patents "per se are not offensive". But the economic interests bound to them should be revealed by the researchers to the public – "also by the DFG". The patent applications made by Oliver Brüstle presuppose "that the target of publicly financed researchers is not merely to develop therapies in an un-selfish manner" (Kollek and Schneider 2001; Sahm 2001; Habermas 2001, p. 37; Montgomery 2001). Patents are rights of use, and rights of use presuppose possession of stem cell lines and thereby prior possession also of embryos. This results in the fundamental question of whether the possession of embryos is possible.

Claims of proprietary entitlements can include goods and services, pension claims, copyrights, patents, and similar things but never human beings. To raise a possession claim concerning a human means to enslave him or her. “A right of freedom never gives mastery over others” (Kirchhof 2002, p. 9; Honecker 1995). The right of free development of personality – for example of the researcher – has its limits at the rights of others (Art. 2, par. 2 GG). “Constitutional freedom means self-determination but not determination over others. It ends in front of the constitutional freedom of the other and does not give any right over his person” (Isensee 2002). Nevertheless, for many centuries slavery was accepted by society – also among people within Christian culture, although already the Apostle Paul had put the axe to its roots when he in 55 after Christ asked the slave master Philemon to treat the slave Onesimus who had escaped from Philemon had become Christian and had been sent back to him by Paulus not as his possession but as his brother (Phlm. 16). Today, at least in most western civilizations, slavery seems to be overcome. In any case nobody talks about slaves when we deal with questions of property.

But the problem of proprietary entitlements of humans gets a new topicality in the course of debate about research with embryonic stem cells. The requirement to be allowed to use the so-called “orphaned” or “surplus” embryos for stem cell research implies using them as a raw material resource for the development of new therapies for hitherto incurable diseases. So this entails regarding embryos neither as persons nor as legal subjects but as things, and therefore as legal objects. To argue in favor of the legitimacy of this requirement, people introduce several distinctions which contend that the embryo (in particular, the embryo in cyro-conservation) is not yet a person and so is not affected by the guarantee of human dignity in Art. 1, par. 1 GG. Among these distinctions belong those between humans, such as growing humans on the one hand versus human life on the other hand, between the embryo in vitro and the embryo in utero, between the pre-embryo and the embryo, the abstract versus the concrete possibility to become a human being, or between embryos living in a communicative environment (possessing interests and having feelings) and embryos where all these characteristics are missing. These distinctions serve the objective of denying the status of being a person and the protection of dignity so the embryo is accorded a lower status, allowing proprietary claims to be made by society for projects of research and therapy. They are reminiscent of attempts to prove the legitimacy of slavery by subdividing people into two types, into those capable of undertaking leadership functions versus those only capable of executing more menial tasks. This attempt at explanation conflicts not only with the Christian conception of man but also the UN-Declaration of Human Rights and the Constitutions of a lot of western countries, which include a guarantee of human dignity. This attempt at explanation is as unconvincing as the judgement of the Supreme Court of North Carolina, who in 1829 refused the claim against a slave–owner due to grievous bodily harm against his slave – he shot her down during an attempt to escape – by explaining that the slave was his property. The act could not be judged as bodily harm because she was completely at the mercy of his powers of disposal (Supreme Court of North Carolina 1829).

In the face of the disposal claims of science and medicine for the so-called orphaned or especially for research and therapy purposes created embryos we have to ask the question of whether society is at the point of turning these embryos into slaves of the twenty-first century. In one sense, the slaves of the Roman Empire, at least in the second century, were in a better situation. It is true that they were property, traded goods, means of production and capital investment but the powers of disposal of the slave masters were limited insofar as they had no rights over the slave's life. So, embryos in vitro have even less rights than slaves. Under the disposal claims of embryonic stem cell researchers, they are reduced to the status of traded goods and a means of production. The German embryo protection law from the 13th of December 1990 sets in § 2 "improper use of human embryos" under pain of punishment. By refusing the trade as well as the research or the therapeutic utilization, the law makes clear that they must not be regarded as a thing. Their creation is only admissible to procure a pregnancy. But the guns to dismantle this bastion for protection of life had been put into position long ago by the Federal Government with the help of the DFG and some people writing comments to the Art. 1 GG. So, a discussion about property as a concept of order cannot be ignored the stem cell debate. It must be shown that a proprietary view of embryos cannot be possible and not of embryos in vitro either, because they do not develop into a person but as a person (BVerGE 88).

Already Immanuel Kant (1724–1804) gave a decisive argument in his Metaphysics of morals 1798. By the act of procreation it is a person who is created and not a thing. So, parents "cannot destroy their child like their creation and their proprietary or leave him to chance, because not only a being of the world but also a citizen of the world was dragged into a position that cannot be indifferent to them according to legal conceptions" (Kant 1798). The same is valid for the nasciturus rule and the embryo in vitro. They are never things for which proprietary claims can be raised but persons with the claim of absolute acceptance. Things never can become persons. "Persons are free beings and thus subjects of a right to justification." (Spaemann 2009) They are not part of a natural species which can be identified by a description or by the attribution of certain features. Right from the start they have a right to recognition. "Person" is not the name of a certain class of beings, but a "nomen dignitatis." "To call somebody a person, means to accord him a certain status, namely the status of being an end in himself." (Spaemann 2009, p. 39) in the debate about research on stem cells Robert Spaemann has consistently emphasized the person's right to recognition by calling to mind not only Kant's categorical imperative but also Thomas Aquinas, who characterized a human being as inherently free and existing for its own sake (STh II-II, 64, 2 ad 3: "Homo est naturaliter liber et propter seipsum existens"). Kant's categorical imperative, which says: "Act in such a way that you treat humanity, whether in your own person or in the person of any other, never merely as a means to an end, but always at the same time as an end," does not prohibit an instrumentalization of human beings.

We could not even exist if we did not treat each other permanently as means to our ends. But this instrumentalization, argues Robert Spaemann, "always has to be reciprocal. We must not reduce other persons to a mere status of being means to our ends" (Spaemann 2009, p. 39). If being a person depended on certain properties like self-consciousness, sensitivity, ability to communicate, autonomy, and so on, the act of recognition would be an "act of co-optation." Persons who join later would depend on the arbitrariness of those who are already recognized, because they could define properties which lead to being co-opted into the community. Such a community would become a club. PGD and gene therapy could be declared terms of admission.

Questions about the status of the embryo – whether the embryo and even the embryo in vitro is already a person – can only be answered in one way. A human being is a person right from the moment of fertilization. "Being a person is not a property, but the being of any human. It starts not later ... than when a new organism, which is not identical with the parents' organism, starts to exist." (Spaemann 2009, p. 42) Spaemann's equation of the moment of fertilization and being a person results from the "impossibility of naming the beginning of a person in time at all" (Spaemann 2009, p. 45). All attempts to separate being a person from the existence of a human organism have proven to be "counterintuitive," which means they are not compatible with current language usage. As an example Spaemann evoke a mother who says to her child "When I was pregnant with you," and who would never say "When there was an organism in me who later has become you" (Spaemann 2009, p. 43). If we abandon the idea of only one criterion for being a person – namely being part of the species *Homo sapiens* and stemming from other members of this species – it will become "a question of power, to which human beings personal rights are awarded and to which not." It is part of human dignity "that he or she is not a co-opted but a born member of the universal community of persons." Thus, if the embryo is his parents' child right from the moment of fertilization and carries in himself or herself the potential of becoming a grown-up person, he or she is a human being that has a right to live. This also applies to embryos in vitro. Also the embryo in vitro carries the potentiality to become a grown-up. Consequently, if there is no development from a "something" to "somebody," there is also no difference ... between an artificial fertilization and the implantation of an artificially created embryo into the uterus ... a "rigid relation of connexity" (Starck 2002a, b). So, the alternative view of "utilization or rejection" (Verwerten oder Verwerfen) Kirchof (2002, p. 30) is an inadequate alternative because it reduces the orphaned embryo to a thing, submitting him or her by means of this to a foreign mastery. If with an uterus an existential condition of his further development is denied to the embryo in vitro, this does not mean that he is passed over to become the property of a Principal Investigator or of society and that he can be subordinated to the interests of research, therapy, or other utilization interests like a thing. Proprietary claims find their inviolable frontier at the human dignity and the right to live for every human.

8.4 A Way Out of the Dilemma Between Research Freedom and Embryo Protection

A solution of the conflict between the freedom of research and the protection of embryos emerges from induced pluripotent stem cells produced from adult somatic cells. Both Shinya Yamanaka and James Thomson produced such stem cells from human cells independently of one another in 2007. With the help of four genes, they reprogrammed mature skin cells such that they resemble embryonic stem cells and therefore have the potential to develop into various, specific types of tissue. "Embryonic potential without the ethical quandary," was the headline used to describe this story of scientific success by the Frankfurter Allgemeine Zeitung, Germany's most competent daily newspaper in bioethics questions (Schwägerl 2007). While Yamanaka's and Thompson's success prompted Ian Wilmut, the creator of the cloned sheep, Dolly, to discontinue research on nuclear transfer as a means of deriving embryonic stem cells (Wilmut 2007) and strengthened the convictions of other stem cell researchers such as Volker Herzog that embryonic stem cell research in Germany has "bet on the wrong horse," (Herzog 2007; Hescheler 2007) other researchers such as Hans Schöler, Oliver Brüstle, Jürgen Hescheler, and Rudolf Jänisch remain convinced that embryonic stem cell research, i.e., ethically unacceptable, embryo-consuming research, is still necessary for the time being. But Hans Schöler does see induced reprogramming as "the method with the greatest potential." (Schöler and Greber 2009).

References

Aquinas Thomas (STh) Summa Theologica II – II, 64, 2 ad 3

Bartram CR (2001) Warum auf den Ethikrat warten? FAZ, 29.6.2001. In: Geyer C (ed) Biopolitik. Frankfurt, 210–212

Beckmann R (2001) Rechtsfragen der PID. Medizinrecht, 19. Jg. 2001: 169–171

Benda E (2001), Verständigungsversuche über die Würde des Menschen. Neue Juristische Wochenschrift, 54. Jg., 2001: 2148.

Böckenförde EW (2003) Dasein um seiner selbst willen. Die Anerkennung der Würde des Menschen, wie sie das Grundgesetz ausspricht, ist auch auf die ersten Anfänge des Lebens zu erstrecken. Deutsches Ärzteblatt 100. Jg., 2003: A 1246–1248

Brüstle O (2000) Gute Nacht, Deutschland? Der Patient ist das Maß aller Dinge: Plädoyer für das Klonen. FAZ 18.8.2000

Brüstle O, Wiestler O (2001) Interview in FAZ 13.6.2001.

Bundesverfassungsgerichtsentscheidungen 88, 203 (252)

Classen CD (2002) Die Forschung mit embryonalen Stammzellen im Spiegel der Grundrechte. Deutsches Verwaltungsblatt 117:144–145

Condic M (2008) When does human life begin? White Paper des Westchester Institute for Ethics & the Human Person. Vol. 1/1

DFG (Deutsche Forschungsgemeinschaft) (2001) Empfehlungen zur Forschung mit menschlichen Stammzellen vom 3.5.2001, Ziffer 8 and 14. FAZ 11.5.2001.

Diedrich K (2001) Plädoyer Pro Embryonenforschung. Forschung und Lehre 6:300

Dürig G (1958) Maunz/Dürig et al. (eds) Grundgesetz-Kommentar Art. 1, Abs. 1, Rn. 28
Fischer J (2001) Pflicht des Lebensschutzes nur für Menschen. Eine theologische Betrachtung der Embryonenforschung. Neue Züricher Zeitung 12.9.2001
Fischer J (2002) Vom Etwas zum Jemand. Warum Embryonenforschung mit dem christlichen Menschenbild vereinbar ist. Zeitzeichen, 3. Jg, Heft 1, 11–13
George RP, Tollefsen C (2008) Embryo. A defense of human life. Doubleday, New York, p 38
Habermas J (2001) Die Zukunft der menschlichen Natur. Auf dem Weg zu einer liberalen Eugenik? Suhrkamp, Frankfurt
Herzog V (2007) Auf das falsche Pferd gesetzt. Interview. Die Tagespost 29.11.2007
Hescheler J (2007) Interview. Die Welt 22.11.2007
Hillgruber C (2002) Die verfassungsrechtliche Problematik der IVF. Zeitschrift für Lebensrecht 11:6–7
Höffe O (2001) Rechtspflichten vor Tugendpflichten. Das Prinzip der Menschenwürde im Zeitalter der Biomedizin. FAZ vom 31.3.2001
Höfling W (2001) Zygote – Mensch – Person. Zum Status des frühen Embryos aus verfassungsrechtlicher Sicht. FAZ vom 10.7.2001; and in: Geyer C (ed) Biopolitik. Frankfurt, 246
Honecker M (1995) Grundriss der Sozialethik. Walter de Gruyter, Berlin, 487
Huber W (2001) Wir stehen nicht erst am Anfang der Diskussion. FAZ 9.8.2001
Isensee J (2002) Der grundrechtliche Status des Embryos. In: Höffe O et al. (eds) Gentechnik und Menschenwürde, Köln, 50
Kant I (1798) Metaphysik der Sitten § 28.In: Weischedel W (ed) Kant I, Werke, Bd. VIII, 394
Kirchhof P (2002) Genforschung und die Freiheit der Wissenschaft. In: Höffe O et al (eds) Gentechnik und Menschenwürde. DuMont, Köln
Koch HG (2001), Unterschiedliches Recht. Zwischen Schwangerschaftsabbruch und Embryonenforschung. Pro-Familia-Magazin 2/2001:10
Kollek R, Schneider I (2001), Verschwiegene Interessen. Die DFG-Position zur Stammzellforschung und der Streit um den Import embryonaler Zellen. Süddeutsche Zeitung 5.7.2001
Laufs A (2000), Soll eine PID eingesetzt werden dürfen? Juristen-Vereinigung Lebensrecht (JVL), Schriftenreihe 17, Köln, 83–84
Markl H (2001a) Der Mensch ist moralisch großzügig geschneidert. Süddeutsche Zeitung, 31.10/1.11.2001
Markl H (2001b) Die vorschnelle Gewissheit der Begriffe. Die Welt, 6.3.2001
Markl H (2002) Schöner neuer Mensch? Piper, München, 45
Montgomery FU (2001) Perspektiven der Fortpflanzungsmedizin – Die vorgeburtliche Diagnostik und das Ethos des Arztes. In: CDU-Landtagsfraktion NRW (ed) Chancen nutzen, Werte achten. Recht und Grenze des gentechnischen Fortschritts, Düsseldorf, 30
Rager G (2004) Der Beginn des individuellen Menschseins aus embryologischer Sicht. Zeitschrift für Lebensrecht 13:66–68
Röger R (2000) Verfassungsrechtliche Grenzen der PID. Juristen-Vereinigung Lebensrecht (JVL), Schriftenreihe 17, Kölner Universitäts-Verlag, Köln, 55–57
Sahm S (2001) Legt offen, was euch bindet. FAZ 15.5.2001
Schäuble W (2001) Vergeßt die Mutter nicht. Mein Ja zur Embryonenforschung. FAZ 21.5.2001
Schockenhoff E (2001a) Einspruch im Namen der Menschenwürde. PID und therapeutisches Klonen instrumentalisieren das menschliche Leben für fremde Zwecke. FAZ 23.4.2001
Schockenhoff E (2001b) Wer ist ein Embryo? Die Politische Meinung 384:13–15
Schöler H, Greber B (2009) Durchbruch in der Stammzellforschung? Die Reprogrammierung von Körperzellen zu pluripotenten Stammzellen. Übersicht und Ausblick. In: Manfred Spieker (ed) Biopolitik. Probleme des Lebensschutzes in der Demokratie. Schöningh, Paderborn, 220
Schwägerl C (2007) Embryonale Potenz ohne ethische Zwickmühle. FAZ 21.11.2007
Spaemann R (1996) Personen. Versuche über den Unterschied zwischen ‚etwas' und, jemand'. Klett-Cotta, Stuttgart, 137
Spaemann R (1999) Die schlechte Lehre vom guten Zweck. FAZ 23.10.1999
Spaemann R (2001) Wer jemand ist, der ist es immer. Es sind nicht die Gesetze, die den Beginn des menschlichen Lebens bestimmen. FAZ 21.3.2001

Spaemann R (2009) Wann beginnt der Mensch Person zu sein? In: Spieker M (ed) Biopolitik Probleme des Lebensschutzes in der Demokratie. Schöningh, Paderborn, 39–50

Spieker M (2008) Kirche und Abtreibung in Deutschland, 2. erw. Aufl. Schöningh, Paderborn, 74–76

Spieker M, Licht und Schatten eines Urteils (1993) Zur Entscheidung des Bundesverfassungsgerichts zu § 218 vom 28.5.1993. In: Thomas H/Kluth W(eds) Das zumutbare Kind. Die zweite Bonner Fristenregelung vor dem Bundesverfassungsgericht, Busse und Seewald, Herford, 317–319

Spieker M (2011) Präimplantationsselektion und Demokratie. Die blinden Flecken der PID-Debatte. Die Neue Ordnung, 65. Jg., 84-99

Starck C (2001) Hört auf, unser Grundgesetz zerreden zu wollen. FAZ 30.5.2001.

Starck C (2002a) Verfassungsrechtliche Grenzen der Biowissenschaft und Fortpflanzungsmedizin. Juristenzeitung 37:1067

Starck C (2002b) Der kleinste Weltbürger. Person, nicht Sache: Der Embryo. FAZ 25.4.2002

Supreme Court of North Carolina (1829) State vs. Mann, 13 N.C. (2 Dev).

Vasella D (2002) Ethos des Heilens contra Würde des Menschen? Ein Diskussionsbeitrag zur Forschung an embryonalen Stammzellen. Neue Züricher Zeitung 9./10.2.2002

Wilmut J (2007) FAZ 27.11.2007

Wolfrum R (2001a) Forschung an humanen Stammzellen: ethische und juristische Grenzen. Aus Politik und Zeitgeschichte 27/2001: 4

Wolfrum R (2001b) Unser Recht auf ein Höchstmaß an Gesundheit. FAZ 29.5.2001

Wolfrum R (2002) Stellungnahme im Hearing des Bundestagsausschusses für Bildung, Forschung und Technikfolgenabschätzung zum Entwurf eines Stammzellgesetzes 11.3.2002, Ausschuss-D

Chapter 9
Interspecies Mixtures and the Status of Humanity

Neville Cobbe

Abstract What does it mean to be a human? Could the mixing of human and nonhuman materials threaten human identity and, if so, how might this happen? This chapter explores such questions in the light of current biological understanding, discussing various features of human life that allegedly mark it off from non-human life, ultimately concluding that any associated characteristics need to be viewed in a holistic manner. The scientific rationale underlying research with various entities that have latterly prompted the greatest controversy is then analyzed in relation to contemporary claims regarding such proposals. Finally, possible approaches towards the ethical evaluation of research involving novel interspecies combinations are presented.

Keywords Chimera • Cybrid • Human identity • Hype • Interspecies

9.1 Introduction

"IS THIS A CLUMP OF CELLS? OR A LIVING BEING WITH A SOUL?"

Amidst the height of heated debate on controversial legislation regarding the proposed use of various interspecies embryos in research, this was the question prominently posed on the front cover of a British newspaper (Laurance 2008). The article began by describing how embryo research had "pitted scientists against bishops" and "divided the country" (including different members of the former Labour Government Cabinet). The aforementioned question, emblazoned in big white capital letters against a dark backdrop that surrounded the magnified image of a pre-implantation embryo, was asserted as being the central issue of the day. It was

---

N. Cobbe
University of Liverpool, The Biosciences Building, Crown Street, Liverpool L69 7ZB, UK
e-mail: Neville.Cobbe@liv.ac.uk

A. Suarez and J. Huarte (eds.), *Is this Cell a Human Being?*,
DOI 10.1007/978-3-642-20772-3_9, © Springer-Verlag Berlin Heidelberg 2011

further claimed that the legislation under debate at the time would challenge "our deepest notion of what it is to be human". If so, how has it done so?

Of course, research that purposefully involves not only the destruction of human embryos but also their production specifically for experimentation had already been sanctioned by statute in the United Kingdom since 1990. With no credible sign of this being subsequently overturned (at the time of the aforementioned debate), it is doubtful that this in itself was the sole cause of newly entrenched division. Instead, the main subject of debate was an interspecies embryo generated by nuclear transfer, in which cells would contain human nuclear DNA but the initial bulk of cellular constituents (including mitochondrial DNA) would be mainly from other species. Despite confusion resulting from hitherto overlapping nomenclature (Cobbe and Wilson 2011, 170–78), the resulting entity gradually came to be commonly described by the term "cytoplasmic hybrid embryo" (St John and Lovell-Badge 2007) or "cybrid embryo" for short. However, the disputed legislation encompassed a raft of additional interspecies entities, each of which was to be covered by the politically concocted moniker of "human admixed embryo" (Homer and Davies 2009). These included a human embryo that has been altered by integration of any non-human DNA sequence into one or more of its cells (in other words, a transgenic human embryo), a human embryo that has been altered by the introduction of one or more non-human cells (a chimeric embryo) and conventional hybrids resulting from fertilization with gametes from different species. The overall aim may have been to ensure that research with any conceivable interspecies embryo containing a significant human component should fall within the remit of the appointed regulatory authority,[1] which had already been responsible for regulating both research and clinical practice with fully human embryos. Nevertheless, the purported justification for generating either "true hybrids" up to 14 days old or transgenic human embryos was not readily apparent in contemporary debates; so the merits of these entities remain especially questionable in the continued absence of any overt demand for their use in sensible research. By contrast, interspecies cybrid embryos had been the subject of intense lobbying, leading to critique not only by clerics (Schofield 2008; Bentley 2008) but also by those advocating the replacement of animals in research (Balls 2008). Moreover, there has been sustained interest regarding the potential ethical questions raised by various chimeric entities, latterly evidenced by consultations on this topic initiated by the Academy of Medical Sciences (2010) in the United Kingdom.

However, what exactly are the novel ethical questions posed by research with human–nonhuman embryonic and foetal mixtures? Could such entities undermine any unique status of our species, perhaps if otherwise viewed as merely "the animal that must recognize itself as human in order to be human" (Agamben 2004, 26), or threaten particular conceptions of human dignity (Schroeder 2008)? If so, how might

---

[1]The Human Fertilisation and Embryology Authority (HFEA), which had ultimately never refused to grant any research license applications.

interspecies mixtures pose any unique challenges? Before we can begin to address these questions, we may need to consider what it is that makes humans unique.

9.2 What Is It that Makes Humans Unique?

You might be inclined to simply say that you know another human when you meet one. However, trying to define specific features of our species that might always set us apart as unique is not as straightforward as it might initially seem. Although various attempts have been made throughout history to distinguish humans on the basis of intellectual capacities for reason and conscience, it is not necessarily clear what distinguishes all members of our species as special on this basis alone. Such a distinction becomes especially problematic in the absence of a radical discontinuity between the mental attributes of humans and those of related species, or when observable differences appear to be primarily quantitative in nature. Even classical concepts of "free will" could arguably be portrayed as a graded series of biological properties possessed to some degree also by invertebrates (Brembs 2011), countering assumptions that humans alone might be capable of decision making. Indeed, most attempts to make qualitative distinctions purely on the basis of isolated human mental faculties may be questionable not just in scientific terms but also from a Judaeo-Christian theological perspective, if the rationality of humans is completely dwarfed by the supreme knowledge of God (Job 38:1–40:5, Isaiah 55:8–9, John 16:30) yet the ingenuity of supposedly lowlier species is praised (Numbers 22:22–33, Proverbs 30:24–28).

Not only is consciousness a feature that is clearly absent from pre-implantation embryos (which have developmental potential for sentience but as yet lack any brain), this may not be a feature that is strictly unique to mature humans either. For example, those under a general anaesthetic temporarily lose self-awareness, whereas self-recognition in a mirror was demonstrated in chimpanzees a number of decades ago (Gallop 1970). Like humans, it appears from a host of psychological experiments that chimpanzees also understand the actions of others in terms of the underlying goals or intentions involved (Call and Tomasello 2008). Moreover, young chimpanzees have latterly been shown to have an extraordinary capability for numerical recollection. In fact, one 7-year-old chimp in particular has been described as performing better than university students tested at a game of memory using the same apparatus and testing procedure (Inoue and Matsuzawa 2007).

Nevertheless, a large cerebral cortex is often considered to be a key feature distinguishing humans from other primates, such that the average human brain at birth is about three times larger than that of our closest primate relatives. However, brain volumes can be reduced to a third of normal in humans with a neurodevelopmental disorder known as "autosomal recessive primary microcephaly" (Ponting and Jackson 2005). Affected individuals have varying degrees of mental retardation (ranging from mild to moderate) and are characterised by a sloping forehead and reduced head size due to considerably reduced brain growth in utero, though their appearance is otherwise normal (Cox et al. 2006). As genes that are mutated in primary microcephaly appear to have been subject to positive selection associated with increasing primate brain size (Montgomery et al. 2011), this has prompted suggestions that adaptive changes in the encoded proteins may have been important factors in the evolution of the human brain. However, caution is needed in extrapolating such findings to defining features of humanity, partly due to the poor correlation between brain size and cognitive abilities in healthy human adults (Tramo et al. 1998) and especially since it would be gravely wrong to suggest that patients with primary microcephaly are anything other than fully human.

Among other morphological transitions from australopithecine ancestors to humans is a reduced reliance on powerful jaw muscles as a means of breaking down food, which is closely correlated with the dramatic increase in cranial capacity. The first of these changes may be associated with a particular member of the class of genes encoding a myosin heavy chain. Intriguingly, although the *MYH16* gene is specifically expressed in the jaw muscles of humans and nonhuman primates, a mutation specifically in the human gene prevents the accumulation of MYH16 protein (Stedman et al. 2004). The apparent inactivation of MYH16 in humans has prompted intriguing speculation that the resulting decrease in jaw muscle size might have removed stress-related constraints on remodelling of the hominid cranium, thereby allowing an increase in brain size (Currie 2004). However, various other genes are also known to have been inactivated since diverging from the chimpanzee lineage (Chou et al. 1998; Winter et al. 2001; Gilad et al. 2003; Wang et al. 2006), and it remains questionable whether apparent losses of function would have necessarily led to the success of humans as a species. Instead, as the genomes of humans and chimpanzees are remarkably similar (Chimpanzee Consortium 2005), many of the visible differences are thought to reflect differential regulation of gene expression (Cáceres et al. 2003).

Meanwhile, human language has been suggested as a unique feature of our species (Anderson 2004), fuelling considerable interest in the FOXP2 transcription factor after a couple of amino acid replacements in the human lineage were found to distinguish this highly conserved protein from that in closely related species (Enard et al. 2002). As mutation of FOXP2 was already associated with severe speech and language difficulties (Lai et al. 2001), it was speculated that the two amino acid replacements described as specific to humans might be involved in the evolution of language. Indeed, the chimpanzee, gorilla, and rhesus macaque FOXP2 proteins are identical to each other and carry just a single difference from the mouse but two differences from the human protein. These substitutions have persisted unchanged in modern humans and appear to have been shared with Neanderthals (Krause et al. 2007). In particular, one of these involved an asparagine-to-serine change that was proposed to create a potential new target site for phosphorylation by protein kinase C (Enard et al. 2002).

However, it is interesting to note that subsequently available data reveal how the aforementioned serine is found not only in humans but is also conserved in dogs, cats, and horseshoe bats – even though the sequence of FOXP2 is relatively diverse amongst echolocating bat species (Li et al. 2007). Moreover, the corresponding residue in FOXP2 from duck-billed platypus (Accession: XP_001509861) is a threonine, and thus conservatively similar. Conversely, although FOXP2 expression appears to be required for learned vocalizations in songbirds (Haesler et al. 2007), conserved sequences from such species (as well as vocal-learning whales and dolphins) do not share the substitutions described as unique to humans (Webb and Zhang 2005). Nevertheless, the threonine-to-asparagine change at position 303 of human FOXP2 is seemingly unique amongst placental mammals, whilst introduction into mice of both FOXP2 substitutions identified in humans has been found to alter their exploratory behaviour and ultrasonic calls (Enard et al. 2009). Moreover, the addition of human versus chimp versions of FOXP2 to neuronal cells otherwise lacking the protein has been shown to differentially affect the expression of multiple genes, which are also differentially expressed in human and chimpanzee brains (Konopka et al. 2009). These studies suggest that at least one of the substitutions affecting FOXP2 in humans may have had profound effects on human evolution, though their precise roles in the development of language remain obscure.

In any event, linguistic impairments had not prevented affected individuals from being recognised as humans (Lai et al. 2001), and it may be questionable to what extent language in humans represents a qualitative difference sharply separating us from other species. Aside from the remarkable ability of Border Collies to learn the meanings of hundreds of words and quickly guess the meaning of a new one (Kaminski et al. 2004; Pilley and Reid 2011), some nonhuman species have been shown to be capable of mastering both semantic and syntactic information in artificially taught languages, notably including a bonobo (Savage-Rumbaugh et al. 1993) and bottlenose dolphins (Herman 1986; Herman and Uyeyama 1999). In addition, our closer ape relatives can communicate context-dependent meaning through extensive use of arm and hand gestures (Cartmill and Byrne 2007; Pollick and de Waal 2007; Genty et al. 2009), which are thought to have served as a stepping stone for the evolution of symbolic communication in language.

So, what is it that makes humans unique? The answer is surely not any one factor in isolation but rather a whole suite of characteristics. Despite the accumulating list of similarities between humans and other species, it is nonetheless clear that differences remain (Premack 2007), albeit more quantitative than qualitative in nature. According to Sarah Chan, "what makes us uniquely human is the ability to shape our own destinies according to our desires" (Chan 2008). At the very least, a long history of interventions from selective breeding (Darwin 1859, 7–43) to "synthetic biology" (Gibson et al. 2010) should make it clear that our species has displayed an unparalleled ability to manipulate the genetics of others. Although expounding a different perspective from that seemingly expressed by Chan, this functional distinction may partly echo theological attempts at defining humanity in terms of the *imago Dei* (Genesis 1:26–28). Rather than naïvely equating God's image simply with a human mind (Editorial 2007), this may be understood as "the potential for authority associated with the human species as a community of stewards and co-creators in relation to God" (Cobbe 2007, 614), whereby we "share in the communion of trinitarian life and in the divine dominion over visible creation" (Commissione Teologica Internazionale 2004, 275). In striving to encompass the totality of humanity, such considerations should be seen as relating to our species as a whole in a relationship and not just to any singular characteristics of isolated individuals.

The degree to which we differ from other animals therefore encompasses everything with which our species has been endowed to manipulate nature and exert authority over the natural world. This can include not just large and intelligent brains (coupled with manual dexterity), but also the ability to cooperate in a society. Although presently hard to account for in terms of simple genetic differences, the remarkable ability of humans to manipulate nature is certainly hard to dispute. Nevertheless, as Andrew Linzey has stressed (Linzey 1994, 54–57, 71), Christian views in particular regarding human dominion over other creatures should emphasize responsibility, following Christ's example of lordship manifest in service (Mark 10:42–45, Philippians 2:5–8). After all, if "we have been created in the image of God, then we have been created to love, because God is love" (Teresa 2002, 83). We may have the power to shape destinies according to our desires (Chan 2008) but with great power comes great responsibility. The questions to which we must therefore now turn are whether the kind of manipulations involved with various interspecies embryos are appropriate with regard to the moral status of such entities, and whether the primary aim underlying such proposals is to seek the good of others.

9.3 Cybrid Embryos: Animal Eggs for Human Cloning?

Long before the immense quantity of women's eggs used in fraudulent human cloning attempts was eventually disclosed (Nature staff 2006), Ian Wilmut suggested that the eggs of other species might be used to reprogram gene expression in the nuclei of adult human cells (Wilmut 2004). After it became clearer how most donated eggs would be wasted in cloning efforts, Stephen Minger argued that it would be more ethical to use eggs from domesticated livestock that are already killed for their meat and can provide a large pool of eggs, rather than encouraging women to risk undergoing invasive procedures for which they receive no direct medical benefit (Minger 2006). Regrettably, some ridiculously irresponsible hype regarding the purported promise of such research was then engineered to combat initial repugnance about crossing species boundaries, yielding multiple press reports suggesting that such research was vital to combat disease and save lives (Fleming 2007; Hammett and Day 2007). Even the leader of the British Conservative Party (in Opposition at the time) was evidently led to believe that treatments for his disabled son might result from interspecies nuclear transfer (Chapman 2008), yet this specific hope was tragically dashed within less than a year when his son passed away.

Notably, Chris Shaw (of King's College London) described previously proposed delays in permitting such experiments as "a real affront to patients who are desperate for therapy" (Henderson 2007), with one of the patients under his care featured in an article claiming that research "using human–animal embryos offers the only glimmer of hope for an otherwise incurable and deadly condition" (Rose 2007). When later questioned about the prospects for such research, he asserted that there was "no obvious biological reason why it should not work" (Science and Technology Committee 2007, Ev 7 [Q49]). In an ensuing letter, he further appealed against closing the door on what he described as "a promising area of research" for those suffering from conditions such as Alzheimer's, Parkinson's, and motor neurone disease, arguing that "research on human–animal hybrids is good news for scientists and patients" (Shaw 2007). Although Shaw himself proposed to use cells derived from interspecies embryos simply for study and had previously indicated that such cells "would not be appropriate for transplantation" (Science and Technology Committee 2007, Ev 7 [Q48]), Lord Walton of Detchant (a Professor of Neurology and former Dean of Medicine) then claimed in Parliament that "the new technique of cloning using the interspecies embryo" would not produce an immunological response. Despite recognised concerns that stem cells derived from components of divergent species may not be tolerated by a human patient's immune system (Wakayama 2004), never mind additional risks, Lord Walton went on to assert:

> Now, if one can use animal cells to produce the type of capsule or framework in which the nucleus from that cell can be implanted, stem cells derived from that cell will be immunologically compatible with the host into whom the subsequent stem cells will be implanted.[2]

Ironically, the absurdly hyped and facile claims promoting such procedures may in turn have helped to fuel further opposition, since many of the strongest hopes and fears alike depended in part on an assumption that viable human-like embryos would be produced (Cobbe 2007, 603). For example, it would seem that the Catholic Bishops' Conference of England and Wales simply took the popularised claims of proponents (Fleming 2007; Henderson 2007; Moss 2007) at face value, consequently arguing that "embryos with a preponderance of human genes should be assumed to be embryonic human beings, and should be treated accordingly" such that interspecies embryos should not be prevented from developing in a woman's womb (Catholic Bishops' Conference 2007). However, it was already doubtful that most such mixed-species cybrid embryos would survive particularly long, since host mitochondrial function is apparently not properly supported by donor nuclei from more distantly related species (Barrientos et al. 1998, 2000; McKenzie et al. 2003). This is especially relevant since oocytes with lower mitochondrial numbers seem to have lower developmental potential (May-Panloup et al. 2005; El Shourbagy et al. 2006), so residual human mitochondria transferred with the donor nucleus would be unlikely to sustain development. Even if sufficient numbers of functioning donor mitochondria could be supplied by other means, further barriers to successful organismal development could be readily envisaged. After all, nuclear transfer with eggs from the same species was already associated

[2]Hansard Official Report (House of Lords), 19th November 2007; Columns 708–09.

with unpredictably variable disruption of gene expression (Humpherys et al. 2002; Somers et al. 2006; Yang et al. 2007). In light of differences in the reprogramming of gene expression during the earliest stages of embryonic development in different mammalian genera (Beaujean et al. 2004; Chen et al. 2006) and some similarities between defects in cloned animals and interspecific hybrids (Vrana et al. 1998; Hiendleder et al. 2004), one would have expected that using eggs from more distantly related species would lead to even more defects in reprogrammed gene expression.

Indeed, most attempts to clone mammals using eggs from distantly related species had only permitted rather limited embryonic development (Dominko et al. 1999; Murakami et al. 2005; Xu et al. 2008). More recently, Robert Lanza and his colleagues reported that embryos fail to grow beyond 16 cells if generated by inserting human nuclei into the eggs of more distantly related species such as cows, mice, or rabbits (Chung et al. 2009). Lanza himself argued that the problem was far more fundamental than just a matter of tweaking the culture conditions (Ledford 2009), going on to describe how "there's no reprogramming towards the embryonic state" such that "these discordant combinations of species are dead on arrival" (Coghlan 2009). Similarly, in an April Fools' Day media announcement, John Burn of Newcastle University declared that embryos produced by inserting human DNA from a skin cell into a cow egg resulted in something that "looks like semolina and it stays like that" but was "never going to be anything other than a pile of cells" (Jha 2008). Notably, development to term was described as the only effective guarantee of complete nuclear reprogramming (Kind and Colman 1999), yet viable cloned offspring had never been obtained when using eggs from distantly related mammals. Therefore, the proposal to use embryonic cells derived in this way to study otherwise unidentified subtle differences associated with late-onset human disease conditions should have seemed all the more ludicrous.

However, skepticism regarding allegedly unique benefits is not in itself sufficient to prohibit research, at least if no obvious harm is apparent and no significant risks are foreseeable. Indeed, it has even been argued that the mismatch between human nuclei and either the reprogramming machinery or mitochondrial proteins of other species ought to be studied (St John et al. 2004); so it could be premature to conclude that experiments involving seemingly useless artefacts such as interspecies "semolina" will no longer be pursued (albeit also assuming that someone would be willing to provide funding). Despite formidable challenges (Beyhan et al. 2007), it was theoretically conceivable that eggs from our closer relatives might be used to clone human or human-like embryos, where eggs from more distantly related mammals (such as cattle) would not suffice and sufficient numbers of human eggs are unavailable. After all, nuclear transfer between closely related mammalian species had been shown to yield at least some viable offspring (Loi et al. 2001; Kim et al. 2007) and other reports of pregnancies (White et al. 1999; Lanza et al. 2000; Sansinena et al. 2005; Yin et al. 2006), indicating that successful reprogramming of gene expression may sometimes be possible using such an approach. The potentially greater feasibility of human cloning using chimpanzee eggs (and the relative improbability of success with those of more distantly related species) may be explored by examining the branch lengths of a phylogenetic tree based on cytochrome *b* (a mitochondrial protein for which sequence data are widely available), revealing that species in which interspecies nuclear transfer has led to pregnancies are generally those separated by comparatively short distances (Fig. 9.1).

As with other mitochondrial proteins (Cobbe and Wilson 2011, 174–75), the distances separating cytochrome *b* in human and chimpanzee are relatively short (albeit longer than in mammalian genera where interspecies nuclear transfer yielded viable offspring), reflecting greater compatibility between human nuclei and chimpanzee mitochondrial function (Kenyon and Moraes 1997). On the other hand, even orangutan mitochondrial and human nuclear genomes apparently encode incompatible oxidative phosphorylation components (Barrientos et al. 2000; Bayona-Bafaluy et al. 2005); so the eggs of species that are far more distantly related (such as cattle) would seem implausible as candidates for human cloning. However, when the use of numerous eggs from women for cloning research was considered unjustifiable because of the risks involved and the lack of personal or guaranteed benefits (Minger 2006), it would be questionable how the use of eggs from non-consenting members of any sentient, intelligent, and endangered nonhuman ape species could be considered any more justifiable. In the United Kingdom, the use of any great apes (such as chimpanzees) in research that might cause suffering had already been effectively prohibited (Home Office 2000), with apparently similar restrictions in the Netherlands and Sweden (Hau and Schapiro 2007). Meanwhile, rights to life and freedom that were previously associated only with humans have been awarded latterly to great apes in Spain (Glendinning 2008). Therefore, assuming that eggs from closely related primates would not be used,

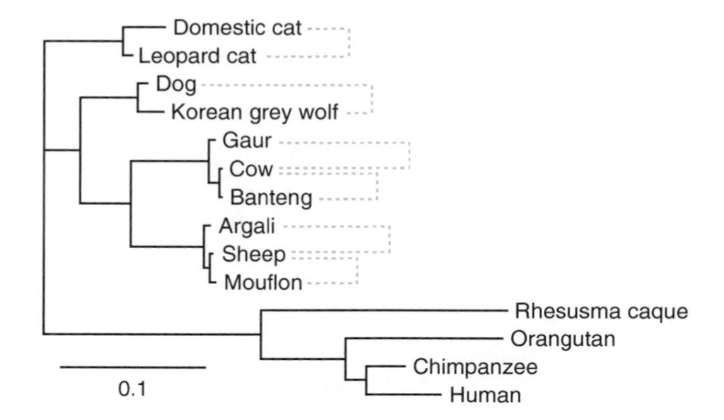

Fig. 9.1 Phylogenetic tree of cytochrome *b* protein sequences, in which the sum of distances along intervening horizontal lines reflects the extent of divergence between protein sequences (the length of vertical lines connecting nodes does not represent actual distances and is merely a reflection of presentation constraints). Pairs of species in which interspecies nuclear transfer reportedly yielded pregnancies or survival to term are connected by *dotted grey lines*. The distance denoted by the *scale bar* (0.1 units) is based on the Whelan and Goldman substitution matrix

there should have been little reason to suspect that the transfer of human nuclei into more readily available eggs from other species would yield viable offspring and thus raise ethical concerns associated with their treatment. As Jason Scott Robert has pointed out (Robert 2006), there may be an important tension when some studies that might be more questionable scientifically could also be those that are less likely to pose significant ethical concerns, and vice versa.

What, then, are the outstanding ethical concerns? Latterly, the Congregation for the Doctrine of the Faith (2008) published a document entitled "Dignitas Personae", which described the use of animal oocytes for reprogramming the nuclei of human somatic cells as "an offense against the dignity of human beings" (n. 33). Interestingly, in considering this to be an offence, no obvious appeal was made to the potential rights or wrongs of destroying the resulting embryo in order to extract stem cells. This markedly contrasts with the description of research that "invariably causes the death" of a living human embryo, which is condemned within the same document as "gravely illicit" due to "the suppression of human lives that are equal in dignity to the lives of other human individuals and to the lives of the researchers themselves" (n. 32). Instead, the use of animal oocytes in place of human oocytes is considered to be "against the dignity of human beings on account of *the admixture of human and animal genetic elements capable of disrupting the specific identity of man*".

Consequently, such objections to interspecies embryos are seemingly due to the mixing of human and nonhuman DNA, yet the moral status of any such entity itself is not directly addressed. Based on the contrasting approach to human embryos, one might infer that the Congregation for the Doctrine of the Faith did not consider that any entities resulting from interspecies nuclear transfer could develop so as to be "equal in dignity to the lives of other human individuals". If so, this conclusion would certainly appear consistent with concurrently available evidence and arguments that such entities would be intrinsically nonviable due to incompatibilities between divergent species (Cobbe 2007, 603–04, 606). However, the question then arises as to how the mixing of human and nonhuman DNA might threaten the dignity of human beings. After all, no obvious reference is specifically made in this regard to previous teaching on related matters, such as the *imago Dei* (Commissione Teologica Internazionale 2004). Although the relevant text of *Dignitas Personae* subsequently mentions unacceptable exposure of humans to unknown health risks associated with any derived stem cells, it is not made clear how such exceptional risks might necessarily arise if transplantation into patients is precluded (notwithstanding Lord Walton's dubious assertions) and research is restricted to controlled laboratory settings. In any event, such risks are seemingly expressed here as a possible additional concern, rather than as the primary objection.

So, how might "the specific identity of man" be uniquely disrupted by such research? In what way does this "represent an offense against the dignity of human beings"? And why was this not previously apparent in response to earlier research where the mixing of human and nonhuman DNA only involved somatic cells (Ruddle and Kucherlapati 1974)? In other words, is there something morally unique about a nonhuman egg that now makes the mixing of human and nonhuman genetic material particularly offensive, even though no sexual act is involved? If so, how is this reflected in attitudes to the use of nonhuman eggs in research more generally? Conversely, if nonhuman eggs have not previously been ascribed a special moral status, and if fusion of somatic cells from different species has not hitherto been considered as unethical, then why should an intrinsically nonviable entity generated from human somatic cells and oocytes from other species be considered as morally unacceptable? Such questions would clearly need to be addressed if it is hoped that the corresponding instruction might be intellectually compelling, rather than appealing to emotional intuition and risking self-contradiction.

Meanwhile, the apocryphal merits of reprogramming human nuclei with the eggs of more distantly related species would be increasingly challenged in the face of stunningly rapid research progress with induced pluripotent stem (iPS) cells (Takahashi and Yamanaka 2006; Stadtfeld and Hochedlinger 2010). Notably, the advent of human iPS cells apparently prompted Ian Wilmut to abandon cloning by nuclear transfer in favour of this direct reprogramming approach (Highfield 2007; Topping 2007), which requires neither eggs nor embryos. On the other hand, those who wished to persist with using cow eggs for human cloning would face increased difficulties in obtaining the necessary funding to compete with an evidently superior alternative (Mayor 2009). Whereas the Association of Medical Research Charities (representing 223 organisations) had intensively lobbied Members of Parliament with a letter claiming that interspecies cloning research "could greatly increase our understanding of serious medical conditions affecting millions of people throughout the UK, and ultimately lead to new treatments" (Laurance 2008), the subsequent lack of any concomitant financial support provides ample reason to question the quality of evidence that supposedly lay behind such beguiling statements.

However, a question may also arise as to whether factors used to generate iPS cells might someday be used in combination with nuclear transfer (Gurdon and Murdoch 2008), in order to increase the fidelity or efficiency of reprogramming steps. This possibility could seem more likely if factor-mediated reprogramming normally requires a gradual process (Stadtfeld et al. 2008), such that low-passage iPS cells generated from adult somatic tissues are less faithfully reprogrammed than embryonic stem cells derived by nuclear transfer (Kim et al. 2010). Slow reprogramming may also be a concern where genomic abnormalities have been reported for various pluripotent stem cells in culture over time (Laurent et al. 2011). Interestingly, elevated expression of the same set of transcription factors (Oct4, Sox2, c-Myc, and Klf4) is known to induce pluripotency in several diverse mammals alike (Shimada et al. 2010; Esteban et al. 2009; Li et al. 2009; Liu et al. 2008; Nakagawa et al. 2008), and serial nuclear transfer is reported to enhance the time available for correct reprogramming during pre-implantation development (Zhang et al. 2007). In theory, one could envisage a stepwise protocol in which reprogramming is initially attempted using factors introduced without viral vectors (Okita et al. 2008; Zhou et al. 2009) and any resulting iPS cells are then injected into enucleated eggs of suitable species for subsequent rounds of cloning by nuclear replacement.

Of course, it is presently hard to see how such experiments in humans could be justified, especially whilst additional technical improvements continue to increase the efficiency of iPS cell production (Lin et al. 2009; Hanna et al. 2009; Zhu et al. 2010; Warren et al. 2010) without any recourse to nuclear transfer. Nevertheless, the demand for women's eggs to use in cloning remains inexplicably strong (BBC News 2007; Maher 2008; Nelson 2009), regardless of its current inefficiency and outstanding ethical concerns (Cobbe 2006; Baylis and McLeod 2007). Consequently, it ultimately remains to be seen whether the generation of iPS cells will necessarily eliminate the demand for additional eggs and embryos in human stem cell research (as one might reasonably hope), or if such advances might also indirectly enhance the feasibility of reproductive cloning (wherever mitochondrial differences are insufficient to preclude viability). If the use of nuclear transfer in humans is not prohibited and if it is also permitted for purposes other than stem cell derivation (Henderson 2008), then this clearly remains an open question.

9.4 Chimeras: Animals with Human Brains or Gametes?

Different ethical issues may be raised by work with some chimeric animals, in which genetically distinct populations of whole cells are derived from human embryos and those of other species. On one hand, if one starts with a human embryo and injects into it pluripotent stem cells from another species, it is doubtful whether the use of such an embryo in research (as ordinarily regulated in various jurisdictions to date) would fundamentally raise any new ethical concerns. Assuming that the resulting interspecies chimera is legally destroyed within 14 days and implantation in a uterus is prohibited, then the most obvious ethical concerns are those that already pertain to destructive human embryo experimentation in general (with attendant questions as to the particular aims and rationale). By contrast, it seems that the reciprocal experiment (whereby pluripotent human stem cells are injected into an embryo of another species) would be subject to different regulatory constraints and might therefore raise novel questions. For example, research in the United Kingdom with various nonhuman species is governed by the Animals (Scientific Procedures) Act 1986, which defines any living mammal (other than humans) as a "protected animal" subject to regulation only after it has developed halfway through the normal period of gestation.[3] This obviously contrasts with the regulation of research with human embryos, which is ostensibly restricted to ensure that embryos do not develop past 14 days or the appearance of the primitive streak.[4] Therefore, it may be possible for a chimeric animal with a significant proportion of human tissue to develop to considerably later stages than would be legally

[3]Sections 1(1) and 1(2)(a) of the Animals (Scientific Procedures) Act 1986 (c. 14).

[4]Sections 3(3)(a) and 3(4) of the Human Fertilisation and Embryology Act 1990 (c. 37).

permissible with human embryos used in research. For those who might otherwise accept research with nonhuman species in preference to analogous experimentation on humans, how should scientists treat a chimeric animal with a significantly human composition?

In light of such questions, the National Academy of Sciences in the USA recommended that research in which human embryonic stem cells are introduced into nonhuman primate embryos should not be permitted for the time being (NAS Guidelines Committee 2005, 40–41). Similarly, in a "Policy Forum" article in the journal *Science*, Ruth Faden and colleagues concluded that it would be unacceptable to graft human neural cells into closely related species at an early developmental stage if the human cells would potentially constitute a large proportion of the host animal's brain (Greene et al. 2005). Nevertheless, the guidelines formulated by such groups are largely voluntary in nature (rather than legally binding) and do not preclude the production of chimeras involving human cells and embryos from either mice or larger nonprimates (so long as such work is subject to careful review). For example, a group of researchers at the Rockefeller University was able to show how human embryonic stem cells could proliferate and integrate after injection into mouse blastocysts when the embryos were cultured in vitro (James et al. 2006). However, after implantation in the uterus of foster mice, the vast majority of these mouse/human chimeric embryos subsequently failed to retain derivatives of the slower dividing human cells or displayed developmental abnormalities. Despite such incompatibilities between the developmental programs of more divergent species, others had shown how injection of human embryonic stem cells into the ventricles of developing embryonic or foetal mouse brains could yield functional human neurons that successfully integrated in the forebrains of mature animals (Muotri et al. 2005). Furthermore, human embryonic stem cells were reported to rapidly produce neurons even in early chicken embryos following transplantation next to partially differentiated tissue (Goldstein et al. 2002).

The potential outcomes of transplanting substantial numbers of human pluripotent stem cells into the developing embryos of larger mammals are far less certain at present. In contrast to the dubious feasibility of interspecies nuclear transfer with more evolutionarily divergent species, conservation in developmental programs appears to be more important than sequence similarity during chimera development. This is vividly illustrated by the successful generation of viable chimeras between pairs of species that are unable to hybridise, including two mouse species (*Mus musculus* and *Apodemus sylvaticus*) with divergent genome sequences (Xiang et al. 2008) as well as interspecific chimeras between mouse and rat (Kobayashi et al. 2010). As Stephen Minger warned (Joint Committee 2007, 177–78 [Q646]), it may be sooner than expected before someone would want to "take human embryonic stem cells and put them into a primate blastocyst and take that blastocyst to mid-gestation or maybe to birth or maybe to 10 years of age". Notably, any relevant concerns about such applications could similarly apply to human stem cells from other sources (Kang et al. 2009; Zhao et al. 2009), not just those derived from embryos.

In addition to issues raised by the potential conferral of human mental traits, there is the possibility of pluripotent human stem cells contributing to the germline of some chimeras, particularly if such cells are incorporated sufficiently early in development. Perhaps due in part to the inefficiency of complete gametogenesis from sources such as embryonic stem cells in vitro (Daley 2007), other species have already been used as surrogates in attempts to make human gametes in vivo (Weissman et al. 1999; Kim et al. 2005; Geens et al. 2006; Wyns et al. 2008). More recently, it has also been suggested how "it may be possible to generate human oocytes from iPS cells in vitro or through human–animal chimeras" (Deng et al. 2011). Although currently an unlikely prospect, further pursuit of such chimeras – possibly involving primates (Lee et al. 2004) rather than mice – might conceivably raise the risk of a human foetus being trapped in the uterus of another species if the resulting creatures are allowed to interbreed, prompting recommendations that this should be prohibited (NAS Guidelines Committee 2005, 39–40). On the other hand, if a chimera with apparently human mental faculties should be either kept in isolation or unable to reproduce with similar creatures because of enforced prohibitions against breeding, this entails automatic denial of what many would otherwise see as fundamental rights in humans.[5] However, before human rights could be deemed as applicable to some interspecies organisms, this arguably might depend on whether the degree of chimerism is considered sufficient to merit human personhood. So, how might one assess the status of a range of interspecies entities containing a significant human component?

9.5 What Is the Status of Interspecies Entities?

Invoking personhood in order to decide on the ethical status of interspecies entities may be fraught with complications, as the concept already has varied definitions. In everyday English, the word "person" has come to mean little more than an individual human being (Thompson 1996; Naffine 2008), yet this in itself does not necessarily mean that nonhumans should be automatically denied personhood (DeGrazia 2007). In fact, the word derives from the Latin *persona*, which originally had the sense of a mask and particular connotations with those worn in Greek and Roman theatres. Building on developments in which the *persona* came to represent a character who acts, Anicius Manlius Severinus Boethius defined a person as "the individual substance of a rational nature" (Boethius [c. 518–521] 1918, 85, 93). However, considerable variation is possible with regard to how the "rational nature" might be understood, with some seeing this as a property of the individual at the present time only, or in terms of future potential, or as a description representative of an entire species.

John Locke defined a person more narrowly as "a thinking intelligent being, that has reason and reflection, and can consider itself as itself, the same thinking thing,

[5]The Universal Declaration of Human Rights (adopted by the General Assembly of the United Nations, 10th December 1948), Article 16.

in different times and places; which it does only by that consciousness which is inseparable from thinking" (Locke [1689] 1825, 225). According to Locke, thinking and consciousness are absolutely essential to the identity of a person, which depends on one's "present sensations and perceptions" (Locke [1689] 1825, 226), such that identity extends back in time only as far as conscious thought. As John Harris pointed out (Harris 1999), all of the elements in Locke's definition (such as intelligence, self-consciousness, memory and foresight) are capacities that exist in degrees, raising the question of whether one can be more or less of a person. Arguing that there is no justification for taking a gradualist approach to personhood or moral status, Harris has instead proposed defining a person as "a creature capable of valuing its own existence". In selecting this definition, Harris acknowledges that this "may include animals, machines, extra-terrestrials, gods, angels and devils" but he also makes it clear that he personally rejects the personhood not only of the embryo and foetus but also of newborn children. However, a definition of personhood premised on evidence of a capacity to value one's existence also excludes a wide range of mature individuals from being considered as persons, including those with potentially suicidal depression.

Therefore, whether or not a particular interspecies combination should be regarded as a person may depend greatly on the chosen definition of personhood. Noting that rationality is a key feature of personhood according to Boethius, Locke and Harris alike, one may choose to focus on the observable or potential cognitive capacities of chimeras containing human neurons in order to evaluate their moral status. In the case of a broader definition as proposed by Boethius, the nature of a chimera containing a significant number of human neurons may be considered earlier in development, whilst development to maturity and evaluation of psychological status throughout life may be required to satisfy the criteria of Locke and Harris. In the case of each definition, the personhood of a cybrid seems extremely unlikely if it lacks any obvious potential to develop to a stage in which rationality might be apparent.

By contrast, Tom Beauchamp has asserted that "no cognitive property or set of such properties confers moral standing" and "that moral standing does not require personhood of any type" (Beauchamp 1999). Describing instead how moral status may depend primarily on a relationship with those who have moral obligations to respect the rights of others, he has emphasized how possession of a right is independent of being in a position to assert it. Nevertheless, Beauchamp also appeals to the moral significance of avoiding pain, suffering, and emotional deprivation, which appears to be as dependent on a complex nervous system as would be the case for cognitive properties. Consequently, it remains doubtful whether an intrinsically nonviable cybrid should have any more significant moral status than a sloughed human skin cell, especially if it has no greater potential capacity for pain or emotion. However, assessing the potential of a mature chimeric organism to experience emotional deprivation would appear no less challenging than predicting its psychological status. If evaluating personhood already appears ambiguous or contentious, then other criteria may need to be considered in order to assess the moral status of interspecies entities.

Alternatively, assuming that recognition of human rights would be restricted to humanity, then could any interspecies entity ever be effectively considered as a fellow human? If one employs Ernst Mayr's biological species concept (Mayr 1996), this implies that various creatures would only belong to our species if they could successfully reproduce with humans on reaching sexual maturity and thereby yield fertile offspring. The presumed lack of viability in interspecies cybrid embryos with primarily bovine mitochondria would seem to rule out their classification as human based on Mayr's definition (Cobbe 2007, 616–617; Cobbe and Wilson 2011, 184). However, if an interspecies entity instead happened to be viable and capable of developing sentience, then this definition could prove to be ethically unacceptable. For example, who would require that any interspecies entity should engage in sexual activity with humans, thereby risking bestiality (or possibly rape), simply for the sake of determining species membership on this basis? Although related use of the evolutionary species concept (Wiley 1978) may rule out the classification of some chimeras as human because of their overall lack of common ancestry, especially if only a relatively small fraction of cells is entirely human, this begs the question of how many cells must be of human origin, and in which bodily organs, in order for a creature to be considered as essentially human.

9.6 Ethical Evaluation of Interspecies Research

A comprehensive view of humanity may reflect particular functional attributes and a particular species membership, but significant moral status is not necessarily limited by either aspect in isolation. Although it had been suggested that ethical difficulties related to cognitive capacity could be avoided by destroying human–nonhuman chimeric embryos at the earliest stages of development (Streiffer 2005), it is doubtful that extraordinary insights would be gained from generating a chimera in which one's capacity to observe the fate of transplanted human cells would be so restricted. Indeed, in a paper describing early development of implanted human–mouse chimeric embryos (James et al. 2006), the authors pointed to the need for more significant contributions of human cells, stressed the need to examine later developmental time points, and concluded by affirming that live chimeric animal models would be a much more valuable research tool. As others have noted, there "may be an irreducible degree of uncertainty about the cognitive nature of the new chimeric animal, and how it would manifest distress, anxiety, or other factors relevant to one's assessment of animal welfare" (Hyun et al. 2007). Such uncertainty could be compounded by our inability to form an adequate conception of the subjective experiences of other species (Nagel 1974). Given such constraints, there may be no way of evaluating a priori the level of humanity of some human–nonhuman chimeric entities with potentially enhanced mental attributes, and hence the extent to which they arguably should be accorded human rights, unless they are permitted to develop to term and mature.

Following Derek Parfit's reasoning (Parfit 1982, 1986, 487–90), if it cannot be shown that it is intrinsically bad for a particular being to ever live, it may be said to benefit such a being by bringing about or preserving its existence once it has been created. The real ethical issues therefore arise not so much in a creature's existence as in its expected treatment, though such issues could overlap if one intentionally engineers an organism with both the potential for significant cognitive status and severely compromised welfare. However, if the moral status of a particular interspecies organism cannot presently be determined without first generating such an entity, this in itself may be sufficient argument against doing so on the basis of a precautionary principle – at least if there is reason to suspect that a creature's welfare might be unjustifiably compromised. Even so, this does not necessarily mean that one should rule out all possible transfer of human cellular material into other species (Karpowicz et al. 2004; Berg 2006; Cobbe 2007; DeGrazia 2007). Rather than advocating either absolute prohibition or acceptance of all conceivable human–nonhuman mixtures, each kind of proposed experiment may need to be evaluated separately in a manner that seeks to uphold both the highest standards of animal welfare and truly beneficial scientific advances.

Mindful of possible tensions between scientific merit and ethical concerns, it might be helpful to explore how potential benefits and risks could be balanced conceptually. For example, the probability of therapeutic benefit arising from research with a particular entity (insofar as this is reasonably foreseeable) could be plotted along one axis of a graph, whilst the likelihood that no harm is caused to any sentient or potentially conscious beings might be plotted along another, as portrayed in Fig. 9.2. Of course, it is in the nature of research that specific outcomes would remain unpredictable beforehand (otherwise the proposed research would be unnecessary), so ordinarily it would be implausible to estimate such probabilities with precision. Nevertheless, even where considerable uncertainty might surround the potential outcomes of experiments involving human–nonhuman chimeras, the proposals under evaluation could be preceded either by analogous experiments involving transplantation of stem cells from other species (NAS Guidelines Committee 2005, 41; Streiffer 2005) or by preliminary experiments that carry less risk of significantly altering higher brain functions (Greely et al. 2007). Thus, the balance of available evidence from prior research could be employed to suggest which proposals would be more likely to lead to clinical benefits than not, as well as whether exceptional harm to the interests of any particular interspecies entity seems relatively unlikely. This sort of scenario would be ideally represented by points in the paler area of the upper right quadrant, signifying research that should be more acceptable ethically. Conversely, ethically impermissible proposals would tend to fall within the darkest regions of the diagonally opposed lower left quadrant, for which the likelihood of harm is greatest yet therapeutic outcomes do not appear feasible.

As Fig. 9.2 illustrates, this kind of ethical analysis entails substantial grey areas, which would be further populated with any increases in the margin of error for likelihood estimates. Shades of grey particularly dominate the remaining quadrants, either where both harm and benefit seem comparatively unlikely (thereby raising

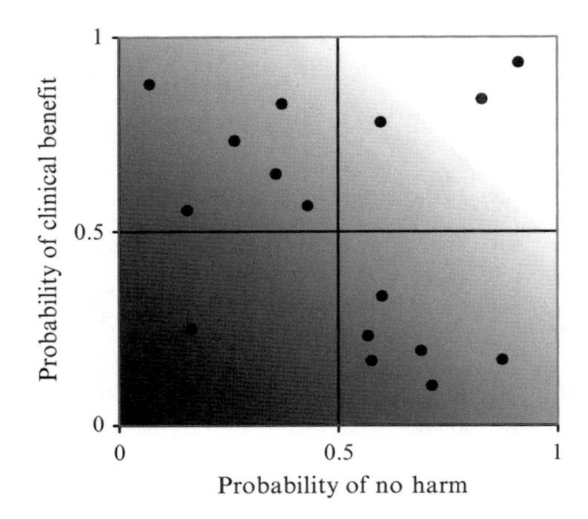

Fig. 9.2 A possible model for ethical evaluation of interspecies research, in which the balance of concerns might be represented graphically. The reasonably foreseeable probability of therapeutic benefit arising from research with a particular entity is plotted along the *Y*-axis, whilst the likelihood that no harm would be caused is plotted along the *X*-axis. Where proposed research appears likely to lead to clinical benefits and the risk of exceptional harm is minimal, this is represented by points in the upper right quadrant. Conversely, ethically impermissible proposals tend to fall within the lower left quadrant, in which the risk of harm is greatest and clinical benefits appear improbable. The *grey areas* include the adjoining quadrants, either where both harm and benefit seem unlikely or where the likelihood of significant clinical benefit needs to be offset by the chance of harm

neither exceptional ethical concerns nor overwhelming justification for pursuit) or where a significant chance of clinical benefit would be offset by the risk of harm. Any judgements regarding such research could then depend on whether greater weight is given to welfare concerns pertaining to possible interspecies creatures or to the value of insights from basic research, even when this has no obvious therapeutic outcome.

However, questions of harm could be extended to include the possibility of emotional harm to surviving individuals in whose interests research would allegedly be conducted, not just potential harm to an interspecies entity that is the subject of research. Having attended or participated in educational events where vulnerable people have been in floods of tears on realising that their hopes of cures were unrealistic, this is assuredly not just a theoretical possibility. A risk of crushing disappointment to patients or their loved ones could have further ramifications by damaging a field of research as trust is ultimately undermined (Blackman 2009). For example, where expectations about the merits of human–cow cybrids had been unreasonably raised by hyped claims and the desperation of sufferers was exploited, the morality of such proposals as portrayed graphically could shift from the lower right quadrant (minimal risk of harm and doubtful chances of benefit) towards the lower left quadrant (increased likelihood of harm but no greater prospect of therapies).

The ethical consequences of such hype could be exacerbated by suggestions that materials to be used in associated research might not necessarily require further specific and explicit consent from donors (Jones and MacKellar 2009). After all, any moral obligation to participate should not only imply fully informed consent but would also depend on a presupposition that research is well founded scientifically (Harris 2005), whereas broad consent is highly contingent on other factors so as to avoid violating the interests of research participants (Broström and Johansson 2011). Notably, the resulting shift in the morality of a research proposal would be independent of the moral status of any actual interspecies entity itself, relating instead to the folly and potential cruelty of unwarranted emotional exploitation or the absence of properly informed consent due to misinformation. Nevertheless, the developmental potential of an interspecies entity would remain relevant to ethical considerations, whether such potential is positively correlated with potential harm (with regard to the welfare of an experimental organism) or negatively correlated (as in the case of grossly misleading hype).

Meanwhile, it may seem disproportionate to express greater alarm about some cross-species mixtures than about extant or potential abuse of recognizably human life. Although particular views of what might be special about humanity could be challenged by various interspecies entities, it is also possible that the underlying assumptions could already be questioned on the basis of existing knowledge about the abilities of other species and their relationship to humans. As many of our traits are shared to some extent with different animals, this challenges the assertions that the generation of interspecies embryos in itself should pose any particularly obvious threat to the status of humanity in general. On the contrary, if we are cumulatively distinguished by a combination of traits that have enabled us to exert authority over other species, then our very ability to engineer novel creatures may be viewed as a potent demonstration of humanity's uniqueness. Instead, a proper evaluation of research with interspecies entities may need to consider the motivations underlying such proposals (insofar as discernible), together with the extent to which these honestly reflect available data and respect the welfare of any resulting creatures according to their developmental potential.

References

Academy of Medical Sciences (2010) Exploring the boundaries: public dialogue on animals containing human material. http://www.acmedsci.ac.uk/index.php?pid=209. Cited Feb 2011

Agamben G (2004) The open: man and animal (trans: Attell K). Stanford University Press, Stanford, CA

Anderson SR (2004) Doctor Dolittle's delusion: animals and the uniqueness of human language. Yale University Press, New Haven, CT

Balls M (2008) The production of admixed animal-human embryos: is it necessary, or merely desirable? Altern Lab Anim 36(2):119–121

Barrientos A, Kenyon L, Moraes CT (1998) Human xenomitochondrial cybrids. Cellular models of mitochondrial complex I deficiency. J Biol Chem 273(23):14210–14217

Barrientos A, Müller S, Dey R, Wienberg J, Moraes CT (2000) Cytochrome *c* oxidase assembly in primates is sensitive to small evolutionary variations in amino acid sequence. Mol Biol Evol 17 (10):1508–1519
Baylis F, McLeod C (2007) The stem cell debate continues: the buying and selling of eggs for research. J Med Ethics 33(12):726–731
Bayona-Bafaluy MP, Müller S, Moraes CT (2005) Fast adaptive coevolution of nuclear and mitochondrial subunits of ATP synthetase in orangutan. Mol Biol Evol 22(3):716–724
BBC News Online (2007) Half-price IVF offered for eggs, 13th September, 2007. http://news.bbc.co.uk/1/hi/england/6992642.stm. Cited Feb 2011
Beauchamp TL (1999) The failure of theories of personhood. Kennedy Inst Ethics J 9(4):309–324
Beaujean N, Hartshorne G, Cavilla J, Taylor J, Gardner J, Wilmut I, Meehan R, Young L (2004) Non-conservation of mammalian preimplantation methylation dynamics. Curr Biol 14(7): R266–R267
Bentley D (2008) Row deepens over hybrid embryo bill. The Irish News, 25th March 2008, p 15
Berg T (2006) Human brain cells in animal brains: philosophical and moral considerations. Natl Cathol Bioeth Q 6(1):89–107
Beyhan Z, Iager AE, Cibelli JB (2007) Interspecies nuclear transfer: implications for embryonic stem cell biology. Cell Stem Cell 1(5):502–512
Blackman S (2009) Promises, Promises. The Scientist 23(11):28–34
Boethius AMS ([c. 518–521] 1918) Contra Eutychen et Nestorium. In: The theological tractates and the consolation of philosophy (trans: Stewart HF, Rand EK). Harvard University Press, Cambridge, MA, pp 72–127
Brembs B (2011) Towards a scientific concept of free will as a biological trait: spontaneous actions and decision-making in invertebrates. Proc R Soc B Biol Sci 278(1707):930–939
Broström L, Johansson M (2011) Broad consent. In: Hug K, Hermerén G (eds) Translational stem cell research: issues beyond the debate on the moral status of the human embryo. Humana Press, New York, pp 237–250
Cáceres M, Lachuer J, Zapala MA, Redmond JC, Kudo L, Geschwind DH, Lockhart DJ, Preuss TM, Barlow C (2003) Elevated gene expression levels distinguish human from non-human primate brains. Proc Natl Acad Sci USA 100(22):13030–13035
Call J, Tomasello M (2008) Does the chimpanzee have a theory of mind? 30 years later. Trends Cogn Sci 12(5):187–192
Cartmill EA, Byrne RW (2007) Orangutans modify their gestural signaling according to their audience's comprehension. Curr Biol 17(15):1345–1348
Catholic Bishops' Conference (2007) Memorandum by the Catholic Bishops' Conference of England and Wales and Linacre Centre for Healthcare Ethics (Ev 87). In: Joint Committee on the Human Tissue and Embryos (Draft) Bill, "Human Tissue and Embryos (Draft) Bill" Session 2006–07, Volume II (Evidence, HL Paper 169-II / HC Paper 630-II). The Stationery Office, London, pp 294–297
Chan S (2008) Humanity 2.0? Enhancement, evolution and the possible futures of humanity. EMBO rep 9(suppl 1):S70–S74
Chapman J (2008) 'Hybrid' hope for my son, by David Cameron. Daily Mail, 21st May 2008, p 5.
Chen T, Zhang YL, Jiang Y, Liu JH, Schatten H, Chen DY, Sun QY (2006) Interspecies nuclear transfer reveals that demethylation of specific repetitive sequences is determined by recipient ooplasm but not by donor intrinsic property in cloned embryos. Mol Reprod Dev 73 (3):313–317
Chimpanzee Sequencing and Analysis Consortium (2005) Initial sequence of the chimpanzee genome and comparison with the human genome. Nature 437(7055):69–87
Chou HH, Takematsu H, Diaz S, Iber J, Nickerson E, Wright KL, Muchmore EA, Nelson DL, Warren ST, Varki A (1998) A mutation in human CMP-sialic acid hydroxylase occurred after the *Homo-Pan* divergence. Proc Natl Acad Sci USA 95(20):11751–11756
Chung Y, Bishop CE, Treff NR, Walker SJ, Sandler VM, Becker S, Klimanskaya I, Wun WS, Dunn R, Hall RM, Su J, Lu SJ, Maserati M, Choi YH, Scott R, Atala A, Dittman R, Lanza R

(2009) Reprogramming of human somatic cells using human and animal oocytes. Cloning Stem Cells 11(2):213–223

Cobbe N (2006) Why the apparent haste to clone humans? J Med Ethics 32(5):298–302

Cobbe N (2007) Cross-species chimeras: exploring a possible Christian perspective. Zygon 42(3):599–628

Cobbe N, Wilson V (2011) Creation of human-animal entities for translational stem cell research: scientific explanation of issues that are often confused. In: Hug K, Hermerén G (eds) Translational stem cell research: issues beyond the debate on the moral status of the human embryo. Humana Press, New York, pp 169–191

Coghlan A (2009) Fatal blow for hybrid embryos? New Sci 201(2694):11

Commissione Teologica Internazionale (2004) COMUNIONE E SERVIZIO: La persona umana creata a immagine di Dio. La Civiltà Cattolica IV:254–286

Congregation for the Doctrine of the Faith (2008) Instruction *Dignitas Personae* on Certain Bioethical Questions. http://www.vatican.va/roman_curia/congregations/cfaith/documents/rc_con_cfaith_doc_20081208_dignitas-personae_en.html. Cited Feb 2011

Cox J, Jackson AP, Bond J, Woods CG (2006) What primary microcephaly can tell us about brain growth. Trends Mol Med 12(8):358–366

Currie P (2004) Human genetics: muscling in on hominid evolution. Nature 428(6981):373–374

Daley GQ (2007) Gametes from embryonic stem cells: a cup half empty or half full? Science 316 (5823):409–410

Darwin C (1859) On the origin of species by means of natural selection, 1st edn. John Murray, London

DeGrazia D (2007) Human-animal chimeras: human dignity, moral status, and species prejudice. Metaphilosophy 38(2–3):309–329

Deng JM, Satoh K, Chang H, Zhang Z, Stewart MD, Wang H, Cooney AJ, Behringer RR (2011) Generation of viable male and female mice from two fathers. Biol Reprod 84(3):613–618

Dominko T, Mitalipova M, Haley B, Beyhan Z, Memili E, McKusick B, First NL (1999) Bovine oocyte cytoplasm supports development of embryos produced by nuclear transfer of somatic cell nuclei from various mammalian species. Biol Reprod 60(6):1496–1502

Editorial (2007) Evolution and the brain. Nature 447(7146):753

El Shourbagy SH, Spikings EC, Freitas M, St John JC (2006) Mitochondria directly influence fertilisation outcome in the pig. Reproduction 131(2):233–245

Enard W, Przeworski M, Fisher SE, Lai CS, Wiebe V, Kitano T, Monaco AP, Paabo S (2002) Molecular evolution of *FOXP2*, a gene involved in speech and language. Nature 418 (6900):869–872

Enard W, Gehre S, Hammerschmidt K, Hölter SM, Blass T, Somel M, Brückner MK, Schreiweis C, Winter C, Sohr R, Becker L, Wiebe V, Nickel B, Giger T, Müller U, Groszer M, Adler T, Aguilar A, Bolle I, Calzada-Wack J, Dalke C, Ehrhardt N, Favor J, Fuchs H, Gailus-Durner V, Hans W, Hölzlwimmer G, Javaheri A, Kalaydjiev S, Kallnik M, Kling E, Kunder S, Moßbrugger I, Naton B, Racz I, Rathkolb B, Rozman J, Schrewe A, Busch DH, Graw J, Ivandic B, Klingenspor M, Klopstock T, Ollert M, Quintanilla-Martinez L, Schulz H, Wolf E, Wurst W, Zimmer A, Fisher SE, Morgenstern R, Arendt T, de Angelis MH, Fischer J, Schwarz J, Pääbo S (2009) A humanized version of Foxp2 affects cortico-basal ganglia circuits in mice. Cell 137(5):961–971

Esteban MA, Xu J, Yang J, Peng M, Qin D, Li W, Jiang Z, Chen J, Deng K, Zhong M, Cai J, Lai L, Pei D (2009) Generation of induced pluripotent stem cell lines from Tibetan miniature pig. J Biol Chem 284(26):17634–17640

Fleming N (2007) Hybrid embryo ban 'would cost lives of patients'. The Daily Telegraph, 5th Jan 2007, p 10.

Gallop GG Jr (1970) Chimpanzees: self-recognition. Science 167(914):86–87

Geens M, De Block G, Goossens E, Frederickx V, Van Steirteghem A, Tournaye H (2006) Spermatogonial survival after grafting human testicular tissue to immunodeficient mice. Hum Reprod 21(2):390–396

Genty E, Breuer T, Hobaiter C, Byrne RW (2009) Gestural communication of the gorilla (*Gorilla gorilla*): repertoire, intentionality and possible origins. Anim Cogn 12(3):527–546
Gibson DG, Glass JI, Lartigue C, Noskov VN, Chuang RY, Algire MA, Benders GA, Montague MG, Ma L, Moodie MM, Merryman C, Vashee S, Krishnakumar R, Assad-Garcia N, Andrews-Pfannkoch C, Denisova EA, Young L, Qi ZQ, Segall-Shapiro TH, Calvey CH, Parmar PP, Hutchison CA 3rd, Smith HO, Venter J (2010) Creation of a bacterial cell controlled by a chemically synthesized genome. Science 329(5987):52–56
Gilad Y, Man O, Paabo S, Lancet D (2003) Human specific loss of olfactory receptor genes. Proc Natl Acad Sci USA 100(6):3324–3327
Glendinning L (2008) Spain: law to enshrine human rights for great apes. The Guardian, 27th June 2008, p 26
Goldstein RS, Drukker M, Reubinoff BE, Benvenisty N (2002) Integration and differentiation of human embryonic stem cells transplanted to the chick embryo. Dev Dyn 225(1):80–86
Greely HT, Cho MK, Hogle LF, Satz DM (2007) Thinking about the human neuron mouse. Am J Bioeth 7(5):27–40
Greene M, Schill K, Takahashi S, Bateman-House A, Beauchamp T, Bok H, Cheney D, Coyle J, Deacon T, Dennett D, Donovan P, Flanagan O, Goldman S, Greely H, Martin L, Miller E, Mueller D, Siegel A, Solter D, Gearhart J, McKhann G, Faden R (2005) Ethics: moral issues of human-non-human primate neural grafting. Science 309(5733):385–386
Gurdon J, Murdoch A (2008) Nuclear transfer and iPS may work best together. Cell Stem Cell 2 (2):135–138
Haesler S, Rochefort C, Georgi B, Licznerski P, Osten P, Scharff C (2007) Incomplete and inaccurate vocal imitation after knockdown of *FoxP2* in songbird basal ganglia nucleus Area X. PLoS Biol 5(12):e321
Hammett A, Day M (2007) Stem cell ban is a death sentence for people like me. The Sun, England, 6th Jan 2007, p 45
Hanna J, Saha K, Pando B, van Zon J, Lengner CJ, Creyghton MP, van Oudenaarden A, Jaenisch R (2009) Direct cell reprogramming is a stochastic process amenable to acceleration. Nature 462 (7273):595–601
Harris J (1999) The concept of the person and the value of life. Kennedy Inst Ethics J 9(4):293–308
Harris J (2005) Scientific research is a moral duty. J Med Ethics 31(4):242–248
Hau J, Schapiro SJ (2007) The welfare of non-human primates. In: Kaliste E (ed) The Welfare of Laboratory Animals. Springer, Dordrecht, pp 291–314
Henderson M (2007) Medicine faces ban on rabbit-human embryos. The Times, 5th Jan 2007, pp 1–2
Henderson M (2008) Loophole in Bill could allow cloning without need for fresh legislation. The Times, 14th June 2008, p 4.
Herman LM (1986) Cognition and language competencies of bottlenosed dolphins. In: Schusterman RJ, Thomas JA, Wood FG (eds) Dolphin cognition and behavior: a comparative approach. Lawrence Erlbaum, Hillsdale, NJ, pp 221–252
Herman LM, Uyeyama RK (1999) The dolphin's grammatical competency: comments on Kako (1999). Anim Learn Behav 27(1):18–23
Hiendleder S, Prelle K, Bruggerhoff K, Reichenbach HD, Wenigerkind H, Bebbere D, Stojkovic M, Müller S, Brem G, Zakhartchenko V, Wolf E (2004) Nuclear-cytoplasmic interactions affect *in utero* developmental capacity, phenotype, and cellular metabolism of bovine nuclear transfer fetuses. Biol Reprod 70(4):1196–1205
Highfield R (2007) Stem cell revolution ends need to use embryos. The Daily Telegraph, 17th Nov 2007, p 1
Home Office (2000) Guidance on the Operation of the Animals (Scientific Procedures) Act 1986. House of Commons Papers No. 321 (Session 1999–2000). Chapter 5, paragraph 5.23: "Other Restrictions". The Stationery Office, London
Homer H, Davies M (2009) The science and ethics of human admixed embryos. Obstet Gynaecol Reprod Med 19(9):235–239

Humpherys D, Eggan K, Akutsu H, Friedman A, Hochedlinger K, Yanagimachi R, Lander ES, Golub TR, Jaenisch R (2002) Abnormal gene expression in cloned mice derived from embryonic stem cell and cumulus cell nuclei. Proc Natl Acad Sci USA 99(20):12889–12894

Hyun I, Taylor P, Testa G, Dickens B, Jung KW, McNab A, Robertson J, Skene L, Zoloth L (2007) Ethical standards for human-to-animal chimera experiments in stem cell research. Cell Stem Cell 1(2):159–163

Inoue S, Matsuzawa T (2007) Working memory of numerals in chimpanzees. Curr Biol 17(23): R1004–R1005

James D, Noggle SA, Swigut T, Brivanlou AH (2006) Contribution of human embryonic stem cells to mouse blastocysts. Dev Biol 295(1):90–102

Jha A (2008) First British human-animal hybrid embryos created by scientists. The Guardian, 2nd Apr 2008, p 4

Joint Committee (2007) Evidence Session, Wednesday 20th June 2007. In: Joint Committee on the Human Tissue and Embryos (Draft) Bill, "Human Tissue and Embryos (Draft) Bill" Session 2006–07, Volume II (Evidence, HL Paper 169-II / HC Paper 630-II). The Stationery Office, London, pp 171–180

Jones DA, MacKellar C (2009) Consent for biobank tissue in somatic-cell nuclear transfer. Lancet 374(9693):861–863

Kaminski J, Call J, Fischer J (2004) Word learning in a domestic dog: evidence for "fast mapping". Science 304(5677):1682–1683

Kang L, Wang J, Zhang Y, Kou Z, Gao S (2009) iPS cells can support full-term development of tetraploid blastocyst-complemented embryos. Cell Stem Cell 5(2):135–138

Karpowicz P, Cohen CB, van der Kooy D (2004) It is ethical to transplant human stem cells into nonhuman embryos. Nat Med 10(4):331–335

Kenyon L, Moraes CT (1997) Expanding the functional human mitochondrial DNA database by the establishment of primate xenomitochondrial cybrids. Proc Natl Acad Sci USA 94(17): 9131–9135

Kim SS, Kang HG, Kim NH, Lee HC, Lee HH (2005) Assessment of the integrity of human oocytes retrieved from cryopreserved ovarian tissue after xenotransplantation. Hum Reprod 20 (9):2502–2508

Kim MK, Jang G, Oh HJ, Yuda F, Kim HJ, Hwang WS, Hossein MS, Kim JJ, Shin NS, Kang SK, Lee BC (2007) Endangered wolves cloned from adult somatic cells. Cloning Stem Cells 9 (1):130–137

Kim K, Doi A, Wen B, Ng K, Zhao R, Cahan P, Kim J, Aryee MJ, Ji H, Ehrlich LI, Yabuuchi A, Takeuchi A, Cunniff KC, Hongguang H, McKinney-Freeman S, Naveiras O, Yoon TJ, Irizarry RA, Jung N, Seita J, Hanna J, Murakami P, Jaenisch R, Weissleder R, Orkin SH, Weissman IL, Feinberg AP, Daley GQ (2010) Epigenetic memory in induced pluripotent stem cells. Nature 467(7313):285–290

Kind A, Colman A (1999) Therapeutic cloning: needs and prospects. Semin Cell Dev Biol 10 (3):279–286

Kobayashi T, Yamaguchi T, Hamanaka S, Kato-Itoh M, Yamazaki Y, Ibata M, Sato H, Lee YS, Usui J, Knisely AS, Hirabayashi M, Nakauchi H (2010) Generation of rat pancreas in mouse by interspecific blastocyst injection of pluripotent stem cells. Cell 142(5):787–799

Konopka G, Bomar JM, Winden K, Coppola G, Jonsson ZO, Gao F, Peng S, Preuss TM, Wohlschlegel JA, Geschwind DH (2009) Human-specific transcriptional regulation of CNS development genes by FOXP2. Nature 462(7270):213–217

Krause J, Lalueza-Fox C, Orlando L, Enard W, Green RE, Burbano HA, Hublin JJ, Hänni C, Fortea J, de la Rasilla M, Bertranpetit J, Rosas A, Pääbo S (2007) The derived FOXP2 variant of modern humans was shared with Neandertals. Curr Biol 17(21):1908–1912

Lai CS, Fisher SE, Hurst JA, Vargha-Khadem F, Monaco AP (2001) A forkhead-domain gene is mutated in a severe speech and language disorder. Nature 413(6855):519–523

Lanza RP, Cibelli JB, Diaz F, Moraes CT, Farin PW, Farin CE, Hammer CJ, West MD, Damiani P (2000) Cloning of an endangered species (*Bos gaurus*) using interspecies nuclear transfer. Cloning 2(2):79–90

Laurance J (2008) Is this a clump of cells? Or a living being with a soul? The Independent, 26th March 2008, p 1ff

Laurent LC, Ulitsky I, Slavin I, Tran H, Schork A, Morey R, Lynch C, Harness JV, Lee S, Barrero MJ, Ku S, Martynova M, Semechkin R, Galat V, Gottesfeld J, Izpisua Belmonte JC, Murry C, Keirstead HS, Park HS, Schmidt U, Laslett AL, Muller FJ, Nievergelt CM, Shamir R, Loring JF (2011) Dynamic changes in the copy number of pluripotency and cell proliferation genes in human ESCs and iPSCs during reprogramming and time in culture. Cell Stem Cell 8 (1):106–118

Ledford H (2009) Hybrid embryos fail to live up to stem-cell hopes. Nature 457(7230):642

Lee DM, Yeoman RR, Battaglia DE, Stouffer RL, Zelinski-Wooten MB, Fanton JW, Wolf DP (2004) Live birth after ovarian tissue transplant. Nature 428(6979):137–138

Li G, Wang J, Rossiter SJ, Jones G, Zhang S (2007) Accelerated *FoxP2* evolution in echolocating bats. PLoS ONE 2(9):e900

Li W, Wei W, Zhu S, Zhu J, Shi Y, Lin T, Hao E, Hayek A, Deng H, Ding S (2009) Generation of rat and human induced pluripotent stem cells by combining genetic reprogramming and chemical inhibitors. Cell Stem Cell 4(1):16–19

Lin T, Ambasudhan R, Yuan X, Li W, Hilcove S, Abujarour R, Lin X, Hahm HS, Hao E, Hayek A, Ding S (2009) A chemical platform for improved induction of human iPSCs. Nat Methods 6 (11):805–808

Linzey A (1994) Animal theology. SCM Press, London

Liu H, Zhu F, Yong J, Zhang P, Hou P, Li H, Jiang W, Cai J, Liu M, Cui K, Qu X, Xiang T, Lu D, Chi X, Gao G, Ji W, Ding M, Deng H (2008) Generation of induced pluripotent stem cells from adult rhesus monkey fibroblasts. Cell Stem Cell 3(6):587–590

Locke J ([1689] 1825) An essay concerning human understanding, 25th edn. Thomas Tegg, London

Loi P, Ptak G, Barboni B, Fulka J Jr, Cappai P, Clinton M (2001) Genetic rescue of an endangered mammal by cross-species nuclear transfer using post-mortem somatic cells. Nat Biotechnol 19 (10):962–964

Maher B (2008) Egg shortage hits race to clone human stem cells. Nature 453(7197):828–829

Mayor S (2009) Lack of funds slows human-animal stem cell research despite legislation in favour. BMJ 338(7688):194–195

May-Panloup P, Chrétien MF, Jacques C, Vasseur C, Malthiery Y, Reynier P (2005) Low oocyte mitochondrial DNA content in ovarian insufficiency. Hum Reprod 20(3):593–597

Mayr E (1996) What is a species, and what is not? Philos Sci 63:262–277

McKenzie M, Chiotis M, Pinkert CA, Trounce IA (2003) Functional respiratory chain analyses in murid xenomitochondrial cybrids expose coevolutionary constraints of cytochrome b and nuclear subunits of complex III. Mol Biol Evol 20(7):1117–1124

Minger S (2006) Junk medicine: therapeutic cloning. The Times, 11th Nov 2006, p 5

Montgomery SH, Capellini I, Venditti C, Barton RA, Mundy NI (2011) Adaptive evolution of four microcephaly genes and the evolution of brain size in anthropoid primates. Mol Biol Evol 28 (1):625–638

Moss L (2007) Scientists say disease research at risk after 'knee-jerk' reaction to embryo proposals. The Scotsman, 5th Jan 2007, p 8

Muotri AR, Nakashima K, Toni N, Sandler VM, Gage FH (2005) Development of functional human embryonic stem cell-derived neurons in mouse brain. Proc Natl Acad Sci USA 102 (51):18644–18648

Murakami M, Otoi T, Wongsrikeao P, Agung B, Sambuu R, Suzuki T (2005) Development of interspecies cloned embryos in yak and dog. Cloning Stem Cells 7(2):77–81

Naffine N (2008) Person. In: Cane P, Conaghan J (eds) The new Oxford companion to law. Oxford University Press, Oxford, pp 885–886

Nagel T (1974) What is it like to be a bat? Philos Rev 83(4):435–450

Nakagawa M, Koyanagi M, Tanabe K, Takahashi K, Ichisaka T, Aoi T, Okita K, Mochiduki Y, Takizawa N, Yamanaka S (2008) Generation of induced pluripotent stem cells without Myc from mouse and human fibroblasts. Nat Biotechnol 26(1):101–106

NAS Guidelines Committee (2005) Guidelines for human embryonic stem cell research. National Academies Press, Washington DC

Nature staff (2006) Korean women launch lawsuit over egg donation. Nature 440(7088):1102

Nelson L (2009) New York State allows payment for egg donations for research. The New York Times, 26th June 2009, p A20

Okita K, Nakagawa M, Hyenjong H, Ichisaka T, Yamanaka S (2008) Generation of mouse induced pluripotent stem cells without viral vectors. Science 322(5903):949–953

Parfit D (1982) Future generations: further problems. Philos Public Aff 11(2):113–172

Parfit D (1986) Reasons and persons. Oxford University Press, Oxford

Pilley JW, Reid AK (2011) Border collie comprehends object names as verbal referents. Behav Process 86(2):184–195

Pollick AS, de Waal FB (2007) Ape gestures and language evolution. Proc Natl Acad Sci USA 104 (19):8184–8189

Ponting C, Jackson AP (2005) Evolution of primary microcephaly genes and the enlargement of primate brains. Curr Opin Genet Dev 15(3):241–248

Premack D (2007) Human and animal cognition: continuity and discontinuity. Proc Natl Acad Sci USA 104(35):13861–13867

Robert JS (2006) The science and ethics of making part-human animals in stem cell biology. FASEB J 20(7):838–845

Rose D (2007) 'Research is now my only hope of a cure'. The Times, 5th Jan 2007, p 7

Ruddle FH, Kucherlapati RS (1974) Hybrid cells and human genes. Sci Am 231(1):36–44

Sansinena MJ, Hylan D, Hebert K, Denniston RS, Godke RA (2005) Banteng (*Bos javanicus*) embryos and pregnancies produced by interspecies nuclear transfer. Theriogenology 63 (4):1081–1091

Savage-Rumbaugh ES, Murphy J, Sevcik RA, Brakke KE, Williams SL, Rumbaugh DM (1993) Language comprehension in ape and child. Monogr Soc Res Child Dev 58(3–4):1–222

Schofield K (2008) MONSTERED; EXCLUSIVE: cardinal attacks 'Frankenstein' embryo plans. Daily Record, 23rd March 2008, p 1

Schroeder D (2008) Dignity: two riddles and four concepts. Camb Q Healthc Ethics 17 (2):230–238

Science and Technology Committee, House of Commons (2007) Government proposals for the regulation of hybrid and chimera embryos. Fifth Report of Session 2006–07, Volume II (Oral and written evidence: HC 272-II). The Stationery Office, London

Shaw C (2007) We must not close the door on victims of terrible diseases. The Daily Telegraph, 18th May 2007, p 8

Shimada H, Nakada A, Hashimoto Y, Shigeno K, Shionoya Y, Nakamura T (2010) Generation of canine induced pluripotent stem cells by retroviral transduction and chemical inhibitors. Mol Reprod Dev 77(1):2

Somers J, Smith C, Donnison M, Wells DN, Henderson H, McLeay L, Pfeffer PL (2006) Gene expression profiling of individual bovine nuclear transfer blastocysts. Reproduction 131 (6):1073–1084

St John J, Lovell-Badge R (2007) Human-animal cytoplasmic hybrid embryos, mitochondria, and an energetic debate. Nat Cell Biol 9(9):988–992

St John JC, Lloyd RE, Bowles EJ, Thomas EC, El Shourbagy S (2004) The consequences of nuclear transfer for mammalian foetal development and offspring survival. A mitochondrial DNA perspective. Reproduction 127(6):631–641

Stadtfeld M, Hochedlinger K (2010) Induced pluripotency: history, mechanisms, and applications. Genes Dev 24(20):2239–2263

Stadtfeld M, Maherali N, Breault DT, Hochedlinger K (2008) Defining molecular cornerstones during fibroblast to iPS cell reprogramming in mouse. Cell Stem Cell 2(3):230–240

Stedman HH, Kozyak BW, Nelson A, Thesier DM, Su LT, Low DW, Bridges CR, Shrager JB, Minugh-Purvis N, Mitchell MA (2004) Myosin gene mutation correlates with anatomical changes in the human lineage. Nature 428(6981):415–418

Streiffer R (2005) At the edge of humanity: human stem cells, chimeras, and moral status. Kennedy Inst Ethics J 15(4):347–370

Takahashi K, Yamanaka S (2006) Induction of pluripotent stem cells from mouse embryonic and adult fibroblast cultures by defined factors. Cell 126(4):663–676

Teresa M (2002) No greater love. New World Library, Novato, CA

Thompson D (ed) (1996) The Oxford compact English dictionary. Oxford University Press, Oxford

Topping A (2007) Professor who created Dolly the sheep to abandon cloning. The Guardian, 17th Nov 2007, p 18

Tramo MJ, Loftus WC, Stukel TA, Green RL, Weaver JB, Gazzaniga MS (1998) Brain size, head size, and intelligence quotient in monozygotic twins. Neurology 50(5):1246–1252

Vrana PB, Guan XJ, Ingram RS, Tilghman SM (1998) Genomic imprinting is disrupted in interspecific *Peromyscus* hybrids. Nat Genet 20:362–365

Wakayama T (2004) On the road to therapeutic cloning. Nat Biotechnol 22(4):399–400

Wang X, Grus WE, Zhang J (2006) Gene losses during human origins. PLoS Biol 4(3):e52

Warren L, Manos PD, Ahfeldt T, Loh YH, Li H, Lau F, Ebina W, Mandal PK, Smith ZD, Meissner A, Daley GQ, Brack AS, Collins JJ, Cowan C, Schlaeger TM, Rossi DJ (2010) Highly efficient reprogramming to pluripotency and directed differentiation of human cells with synthetic modified mRNA. Cell Stem Cell 7(5):618–630

Webb DM, Zhang J (2005) FoxP2 in song-learning birds and vocal-learning mammals. J Hered 96 (3):212–216

Weissman A, Gotlieb L, Colgan T, Jurisicova A, Greenblatt EM, Casper RF (1999) Preliminary experience with subcutaneous human ovarian cortex transplantation in the NOD-SCID mouse. Biol Reprod 60(6):1462–1467

White KL, Bunch TD, Mitalipov S, Reed WA (1999) Establishment of pregnancy after the transfer of nuclear transfer embryos produced from the fusion of argali (*Ovis ammon*) nuclei into domestic sheep (*Ovis aries*) enucleated oocytes. Cloning 1(1):47–54

Wiley EO (1978) The evolutionary species concept reconsidered. Syst Zool 27(1):17–26

Wilmut I (2004) The moral imperative for human cloning. New Sci 181(2435):16–17

Winter H, Langbein L, Krawczak M, Cooper DN, Jave-Suarez LF, Rogers MA, Praetzel S, Heidt PJ, Schweizer J (2001) Human type I hair keratin pseudogene *φhHaA* has functional orthologs in the chimpanzee and gorilla: evidence for recent inactivation of the human gene after the *Pan-Homo* divergence. Hum Genet 108(1):37–42

Wyns C, Van Langendonckt A, Wese FX, Donnez J, Curaba M (2008) Long-term spermatogonial survival in cryopreserved and xenografted immature human testicular tissue. Hum Reprod 23 (11):2402–2414

Xiang AP, Mao FF, Li WQ, Park D, Ma BF, Wang T, Vallender TW, Vallender EJ, Zhang L, Lee J, Waters JA, Zhang XM, Yu XB, Li SN, Lahn BT (2008) Extensive contribution of embryonic stem cells to the development of an evolutionarily divergent host. Hum Mol Genet 17(1):27–37

Xu X, Liu G, Chen J, Chen J, Sha H, Wu Y, Zhang A, Cheng G (2008) Goat MII ooplasts support preimplantation development of embryos cloned from other species. Sheng Wu Gong Cheng Xue Bao 24(3):430–435

Yang J, Yang S, Beaujean N, Niu Y, He X, Xie Y, Tang X, Wang L, Zhou Q, Ji W (2007) Epigenetic marks in cloned rhesus monkey embryos: comparison with counterparts produced *in vitro*. Biol Reprod 76(1):36–42

Yin X, Lee Y, Lee H, Kim N, Kim L, Shin H, Kong I (2006) *In vitro* production and initiation of pregnancies in inter-genus nuclear transfer embryos derived from leopard cat (*Prionailurus bengalensis*) nuclei fused with domestic cat (*Felis silverstris catus*) enucleated oocytes. Theriogenology 66(2):275–282

Zhang AM, Chen JQ, Sha HY, Chen J, Xu XJ, Wu YB, Ge LX, Da HW, Cheng GX (2007) Secondary SCNT doubles the pre-implantation development rate of reconstructed interspecies embryos by using cytoplasts of Sannen dairy goat ova. Fen Zi Xi Bao Sheng Wu Xue Bao 40 (5):323–328

Zhao XY, Li W, Lv Z, Liu L, Tong M, Hai T, Hao J, Guo CL, Ma QW, Wang L, Zeng F, Zhou Q (2009) iPS cells produce viable mice through tetraploid complementation. Nature 461 (7260):86–90

Zhou H, Wu S, Joo JY, Zhu S, Han DW, Lin T, Trauger S, Bien G, Yao S, Zhu Y, Siuzdak G, Schöler HR, Duan L, Ding S (2009) Generation of induced pluripotent stem cells using recombinant proteins. Cell Stem Cell 4(5):381–384

Zhu S, Li W, Zhou H, Wei W, Ambasudhan R, Lin T, Kim J, Zhang K, Ding S (2010) Reprogramming of human primary somatic cells by OCT4 and chemical compounds. Cell Stem Cell 7(6):651–655

Chapter 10
The Boundaries of Humanity: The Ethics of Human–Animal Chimeras in Cloning and Stem Cell Research

William B. Hurlbut

Abstract Advances in molecular and cell biology are opening new possibilities for combining human and animal cells, tissues, and organs. These projects could have important scientific and medical benefits, but open ethical dilemmas that challenge our traditional notions of the boundaries of humanity. Beginning with an overview of the goals and likely limitations of this research, we seek a framework of fundamental principles that can at once open avenues of scientific advance and defend human dignity. Reflection on these difficult dilemmas can promote a deeper understanding and appreciation of what defines and distinguishes the human creature.

Keywords Chimera • Cloning • Embryonic stem cells • Human dignity • Hybrid

10.1 Introduction

In the recently released doctrinal instruction Dignitas Personae, the Vatican decisively condemns *hybrid cloning* as "an offense against the dignity of human beings on account of *the admixture of human and animal genetic elements capable of disrupting the specific identity of man.*" (Congregation for the Doctrine of the Faith 2008)

The history of this proposal provides an important perspective on the deepening dialogue over the ethics of advancing biotechnology. This project, which involves nuclear transfer by the placement of a human somatic cell within the cytoplasm of an enucleated animal egg, was recently endorsed and supported as part of the United Kingdom's new legislation governing reproductive technologies (Fleming 2007;

W.B. Hurlbut
Department of Neurology and Neurological Sciences, Stanford University Medical Center, Stanford, CA, USA
e-mail: ethics@stanford.edu

A. Suarez and J. Huarte (eds.), *Is this Cell a Human Being?*,
DOI 10.1007/978-3-642-20772-3_10,

Hopkin 2007). It was intended to bypass the controversies and practical difficulties of obtaining human oocytes for research cloning for the production of human embryonic stem cells. Nonetheless, in a society where the creation and destruction of fully natural human embryos has met with little objection, the admixture of human and animal elements provoked a storm of public criticism. This reaction provides important insights into the contours of the moral mind, and serves as a warning that the cold calculation of simple linear logic will not be adequate to anticipate the human meaning of our biotechnological projects.

If we consider the range of possible uses of human–animal hybrids and chimeras, we are met with a perplexing mix of potentially valuable scientific projects and difficult ethical dilemmas that challenge our traditional notions of the boundaries of humanity. Reflection on these possibilities can clarify legitimate domains of science and at the same time promote a deeper understanding and appreciation of what defines and distinguishes the human creature.

In what follows I will provide a brief outline of possible scientific projects employing human–animal hybridization and chimerization in cloning and stem cell research. First, I will describe the scientific goals, including certain likely limitations of these projects, followed by a discussion of their ethical implications. I will then propose some general principles within which to frame our reflection, and conclude with a series of specific recommendations for setting public policy governing chimeric cloning and stem cell biology. These recommendations are put forward, not in a spirit of absolute advocacy, but as preliminary provisions intended to provoke discussion and further reflection.

10.2 Scientific Projects

Genetics: Using techniques of recombinant DNA, gene knockouts and siRNA, genes can be added, deleted or silenced. Human stem cells carrying these alterations could be incorporated into developing animal embryos to allow the production of human proteins, enhanced cells with therapeutic powers and possibly immune compatible tissues and organs.

Already there are many scientific and commercial projects involving transgenic animals carrying human genes. Stem cell biology may add a wider range of possibilities with a deeper integration of human and animal characteristics. Currently, there are projects to remove surface antigens of animal cells (or replace them with human surface markers) to allow the abundant animal production of histocompatible tissues and organs for transplantation into human patients (Dekel et al. 2002). While this does not have the immediate ethical implications associated with alterations of form or function, it does raise questions about the importance of species boundaries and the defensive biological "distancing" that prevents infectious microbial diseases and pernicious parasitic colonization, and that provides barriers to interspecies reproduction.

Cognate genes could be selectively interchanged to allow study of distinctive proteins and their implications for qualitative and chronological differences in human and animal species development (Enard et al. 2002). Human cells with programmed reporter genes such as green florescent protein (GFP) could be used to track cell migration and differentiation within animal embryos.

Through fusion of cells it is possible to create human–animal synkaryons and by provoking asymmetric segregation of chromosomes during cell division of these synkaryons it is possible to transfer whole chromosomes between mammalian species or create cells lacking a specific chromosome or carrying an extra copy of a specific chromosome. A Japanese line of mouse cells carries an entire human chromosome (Suzuki et al. 1997). This could allow the creation of animal models of human aneuploid pathologies (such as Down syndrome) and provide new systems for the testing of drugs (Miller 2005).

Since most species barriers seem to be the result of surface receptor blockades to sperm penetration, techniques of genetic alteration might allow animal models for the study of human fertilization and possibly research on the role of species differences in genetic imprinting during early development.

Whole genetic cassettes for dimensions of differentiation and morphogenesis (together with switches to turn them on while turning off the natural programs of their animal host) could be incorporated into stem cells which could then be aggregated into developing embryos allowing their selective expression in specific body segments, biological compartments, or developmental modules.

Such targeted replacement of specific genes or larger genetic developmental programs may allow interspecies transfer of structural patterns such as human limbs, laryngeal structures, or maybe even upright posture or human facial features. Likewise, it might be possible to confer to another creature certain human neurological forms such as an expanded forebrain, human-like cortical cell layering, increased cortical folding (sulci and gyri), or characteristic patterns of synaptic proliferation and connection (Emes et al. 2008; Molnár et al. 2006). In a similar manner, there are animal genes (or developmental programs) that might be placed in the human genome to endow desired characteristics such as enhanced musculature, increased bone strength, or changes in sensory discrimination such as increased powers of visual apprehension or acuity (Savulescu 2003).

Given the pleiotropic nature of most genes and the polygenetic character of most traits, as well as the complex spatial and temporal dynamics of development, it seems unlikely that simple genetic transformations would permit the integrated organismal growth necessary to produce viable creatures with dramatic ambiguities at the boundaries of humanity. Nonetheless, we know that the phylogenetic process that led to the human species must have contained many intermediate forms. Some have suggested such "near-humans" might be useful for scientific study, for organ harvesting, and for specialized roles such as space exploration or as commercially profitable "curiosities" or "objects of art" (Rose 1997).

Cells, Tissues and Organs: Interventions in developmental processes through infusion or transplantation of cells, tissues or whole organic parts, could create animal forms that challenge our intuitive boundaries of humanity.

Aggregating human and animal blastomeres or infusing human ES cells into developing animals will establish specific human cell lineages and clusters within chimeric creatures (Almeida-Porada et al. 2004; Chamberlain et al. 2007). The human-chimp equivalent of the "geep" (goat and sheep), formed by the interspecific fusion of blastomeres has already been proposed (Weiss 2005; Anonymous 1984).

Studies with SCID (severe combined immunodeficiency) mice have established that transplanted parts of human embryos and fetuses can integrate and develop within neonatal and adult SCID animals. Stanford stem cell biologist Irving Weissman has infused human neural progenitor cells into the intraventricular zone of newborn SCID mice and they have migrated, differentiated, and (apparently) functionally integrated into the mouse brains (Tamaki et al. 2002; Greely et al. 2007). Likewise, fetal organ primordia for human kidneys and hearts, limb buds for human hands, and cellular modules for whole human bones (including a two centimeter Tibia), as well as a functional immune system, have been placed in SCID mice and developed in form and function (Weissman I, personal communication). This raises the strange question of what (or how many) human parts might be grown in a larger and longer lived SCID animal for production of later transplantable organs – kidneys, livers, brain tissues, sensory organs, or even genitalia? Clearly, successful production of human embryos with somatic cell nuclear transfer (SCNT), followed by implantation and fetal harvesting, would open avenues toward such morally challenging prospects.

Success in xenographic transplantation of adult gonadal tissues suggests earlier (and thus immunologically invisible) fusions during embryogenesis might also allow abundant cross-species gamete production (Dobrinski et al. 2000).

Le Douarin and Balaban have done a series of dramatic experiments in which they transplanted whole clumps of primordial neurological tissues of early embryos between quail and chicks. These transplants resulted in cross-species transfer of species-typical motor programs such as the posture and vocal pattern of crowing (Travis 1997). In another more recent series of embryonic neural transplants, Balaban demonstrated that "an inborn auditory perceptual predisposition is transferable between species" (Long et al. 2001). Thus, it might be possible to fuse portions of human embryonic neural tissues into non-human primates (or other large mammals) with transfer of certain perceptual, cognitive, and motor characteristics (Shreeve 2005).

Terrance Deacon has studied rat brains transplanted with human fetal neuroblast cells from the midbrain and the striatum and found that the transplanted cells exhibited the "time clock" of the donor animal but their patterns of differentiation were directed by their context within the host. The resulting heterochrony caused asynchronous growth and disruptions of integrated developmental organization (Deacon 1990). He reports that these qualities of intrinsic chronology characterize fused blastomeres as well. Yet, interventions early in development would be essential for the formation of human-like structural architecture or functional features in chimeric creatures. This implies that we may not be able to effectively substitute or integrate embryonic and fetal cells except from closely related species with similar spatio-temporal developmental patterns.

Each of the above-mentioned projects could provide important insights in basic biology or organic materials for medical therapies. However, notwithstanding their almost science fiction implications, the success and human significance of these experiments are likely to be less dramatic than their description might imply. An organism is a living whole, a complex, dynamic network of interdependent and integrated parts. Incompletely constituted or severed from their source in which they are a natural coherent part, embryonic fragments may have a certain developmental momentum, but ultimately they become uncoordinated and disorganized growth.

A natural organism, on the other hand, is a self-sustaining and harmonious whole, a unified being with an inherent principle of organization that orders and guides its development (Hurlbut 2005). Between even closely related species there may be difference of timing in development through subtle differences in epigenetic factors and cell cycle synchronies. Highly complex spatio-temporal dynamics play a major role in the development of an organism, including quantitative factors in gene expression and diffusion patterns of crucial morphogens.

Moreover, organisms cannot be created along an infinite gradient of change, there is no simple gradualism; the similarity and separation of species is more like a Chinese checkers board where certain combinations of organic parts and powers are compossible, but others simply do not coordinate sufficiently to allow natural growth, development, or even ongoing survival. Nonetheless, some of these projects may produce possibilities that will infringe on our sense of human identity and distinctive dignity amid the categories within which we comprehend our place in the order of the natural world.

10.3 Principles

In proposing the essential philosophical tools with which to consider the ethical implications of these new powers, bioethicist Cynthia Cohen has suggested principles and practical constraints with which we may proceed (Karpowicz et al. 2004, 2005). These include the importance of identifying the defining features and functions that distinguish humans from other animals, deciding what percentage of these characteristics would blur the species boundaries, and discerning what stage of commingling of human genes or cells would transfer distinctly evident human anatomy or neurological architecture. While I agree with this approach, and much of her moral analysis, I want to comment on three of her "rejected" categories of consideration: teleological philosophy, taboos and the concept of kinds. These reflections will, in turn, prepare us to discuss the more central considerations in the definition of the boundaries of humanity and the ethics of human–animal chimerization.

First, contrary to Dr. Cohen's opinion, I believe that "to intervene into [the normal functions of organisms] or to keep them from reaching their usual biological ends," *is* a fundamental moral concern when dealing with the human organism

(Karpowicz et al. 2005). Human beings are, as a species, *sui generis*, incommensurate for the purposes of ethical analysis. Human embryos, even at the earliest stages of development, must never be used instrumentally as mere biological resources in our scientific project. The value of an individual human life cannot be outweighed by other goods in a purely utilitarian calculus – most specifically for the benefit of advances in biomedical science. Otherwise, we degrade the very humanity we are trying to heal.

Second, regarding taboos that replace reason with an irrational power of prohibition, clearly superstitions can be "corrected" and social prejudice can be "educated out" by honest information and genuine enlightenment. But, contrary to Cohen, the source of all taboos is not simply cultural convention; there are pre-social psychological responses that seem evident as human universals or near-universals (Brown 1991).

The rightful source of our moral awareness (and sense of species defining distinctions) is not mere ratiocination; fundamental emotions and basic connections within the relational dynamics of life give certain pre-rational human responses a more primary and privileged position, whether we like it or not. These relate preferentially to aspects of our lives that are deeply connected with crucial life transitions and biological imperatives: most specifically, reproduction, nourishment, and death. We should not denigrate or seek to rationally reduce or wholly override these natural precautionary concerns. For example, however irrational it may seem by objective scientific analysis, we should keep human genes and human cells out of the lineage of our food stocks. Likewise, we should take seriously the natural human repugnance to mixing human and animal species at the most fundamental levels of reproduction and development. Revulsion and disgust, "creepiness" and "weirdness"; these are telling us something. Nonetheless, we should seek coherent, consistent, and broadly accessible reasons for our ethical judgments. And, for serious purposes we may at times override our natural moral intuitions and sentiments in fashioning and enforcing science policy positions.

Finally, and I think most importantly, regarding natural categories and kinds we should not be too quick to overextend our insights from evolutionary gradualism to think of all species boundaries as fluid and all essentialism as outdated "folk psychology." Professor Cohen's statement that "the biological categorization of species is empirical and pragmatic, a constantly developing effort that has little to do with moral judgments," could be turned back on *us* as we seek to define the boundaries of human dignity (Karpowicz et al. 2005). Heightened perceptual discriminations and distinct conceptual categories are an essential element of our natural mental capacities and are crucial for constructing an ordered and intelligible world that allows both coherent comprehension and genuine interpersonal communication. In many areas of perception such as apprehension of color and phonemic discernments, we have heightened discrimination at category borders (like the seven colors in what is actually a continuous gradation of light wavelengths in the rainbow). In the same way, our minds have evolved to discern distinctions of significance and to form conceptual categories that order and organize our understanding – and these have crucial significance in our psychological and social functioning. Most central for our current considerations, this applies to parts as well as wholes: I disagree that psychological "emergent human mental capacities" or "supercellular psychological human functions" and their associated faculties are the only grounds that signal the human kind (Karpowicz et al. 2004, 2005). There are other features of the human body whose species distinguishing character we need to protect – faces, hands, and voices to name a few. To sever a sign from its source is to debase the coinage, to draw down on the capital of truth and thereby erode the intelligibility of the order of being. When we can no longer read the meaning of a human message as though it flows forth from a human source in its deeply personal existential embodiment, we will have finally and ultimately dissolved the boundaries of humanity and squandered the very riches of the mysteriously intelligible moral order of the created world. We are, to ourselves and others, only understandable as a psycho-physical unity; as Benedict Ashley says, we are "embodied intelligent freedom" (Ashley et al. 2006).

Nonetheless, I do agree with Professor Cohen that the emergence of our higher human functions is a highly contingent process. Individual development involves a fantastic choreography of physical form and personal experience, and partial expressions of our personhood are unlikely to be easily transferred within simple organic units as though they are independently identifiable pieces of a larger puzzle. There may be no realistic danger of edging out or transferring our distinctive human identity in parts or pieces.

10.4 Defining the Human Organism

With these reflections in place, we can now approach the central question of what defines a human organism, and, most specifically, how biomedical research with human–animal chimeras might proceed in a way that at once advances scientific knowledge while defending the boundaries of human dignity.

We begin with an affirmation of the inviolability of human life from fertilization to natural death. Every embryo, as a unique and unrepeatable gift of life, carries forward from its inception a natural continuity of human identity across all phases and stages of its development. By its very nature, an embryo is a developing being. Its wholeness is defined by both its manifest expression and its latent potential; it is the phase of human life in which the "whole" (as the unified organismal principle of growth) precedes and produces its organic parts. The philosopher Robert Joyce explains: "Living beings come into existence all at once and then gradually unfold to themselves and to the world what they already but only incipiently are." To be a human organism is to be a whole living member of the species *Homo sapiens*, with a human present and a human future evident in the intrinsic potential for the manifestation of the species typical form. Joyce continues: "No living being can become anything other than what it already essentially is" (Joyce 1978).

With the new perspective of Systems Biology we are gaining a greater appreciation of the natural basis for the emergent powers of living wholes, and the crucial developmental differences that distinguish distinct species of organisms. The very word organism implies organization, an overarching principle of unity, a cooperative interaction of interdependent parts subordinated to the good of the whole. As a living being, an organism is an integrated, self-developing, and self-maintaining unity under the governance of an immanent plan. Such a conception of the biological organism adds the understanding that a living being is not merely a mechanism but rather, a dynamic system, an interactive web of interdependent processes that expresses emergent properties not predictable from the biochemical parts.

Indeed, in defining those qualities that characterize human nature and its distinctive dignity, it is crucial to acknowledge the seamless unity of human life across all of its stages. Our distinguishing capacities are already present in the single cell embryo as a kind of "harnessed emergence," an intrinsic potency of form. From a scientific perspective, there is no meaningful moment when one can definitively designate the biological origins of a defining human characteristic such as consciousness. The human being is an inseparable psychophysical unity. Our thinking is in and through our bodily being, and thus the roots of our consciousness reach deep into our development. The earliest stages of human development serve as the indispensable and enduring foundations for the powers of freedom and self-awareness that reach their fullest expression in the adult form.

Moreover, there is a unity of nature in the manifold expression of the human form. This is evident if we pause to ponder the diverse (and often amusing) proposals for the single feature that distinguishes the human being. We have been variably described as the "featherless biped," the "laughing animal," the "only animal that cooks," the "political animal," the "animal capable of speech, memory and imagination," and the "moral animal" (Thomas 1984). Each of these carries an element of truth but fails to fully define either our uniqueness or the breadth of our distinctive powers. A more solid root of thought, tracing back to Aristotle, locates our distinguishing identity in our distinctive capacity for deliberation and speculative reason, and is generally summed up in the concept of the "rational animal." Clearly, our capacity for comprehensive understanding and control of the world is central to human identity and dignity. We are the creature that sees more than the pixels, we recognize the pattern and "get the picture." We alone seem to penetrate to the core insight that we exist within an ordered world, a cosmos in which the material and the moral flow forth from a single creative source.

When we step back and seek to understand the biological foundations of our rational nature, we recognize again that such a nature is grounded in the full continuity and coherence of our developmental origins and manifest being. We are a psychophysical unity: "embodied, intelligent freedom." Our highest capacities of mind are inseparable from our physical form. With our upright posture we are the "beholding creature" with a comprehending gaze toward distant horizons. Our unique combination of senses and our perceptual range cross-complement one another to allow polymodal synthesis, multidimensional understanding, and metaphor. Above all, sight allows insight and with insight comes the conceptual categories that organize our understanding and provide the foundations of imagination and creative recombination. These, in turn, empower a cognitive drive that culminates in a "cosmological imperative," the quest for coherent comprehension, ideals, and moral meaning.

Our defining cognitive capacities are in turn inseparable from the rational action expressed through the range of motion of our body, most specifically the fine motor control of our arms and hands (which Aristotle called the "tool of tools"). These human features provide the creative powers to implement into material form the organized images from mental templates. More fundamentally, however, recent studies in developmental psychology suggest that physical motion provides a crucial conceptual infrastructure for memory and mental operation. What we think of as "creativity" may be inseparably a cognitive and motor capacity. Moreover, our posture, gait, and subtle flexibility of facial expression (including the fine motor control of phonation) provide the essential elements of our unique capacities for complex communication, empathy, and genuine intersubjectivity (Rowe and Goldin-Meadow 2009). Together, these natural proclivities for community, cultural progress, and shared spiritual unity are essential to our definition as beings of a rational nature.

10.5 Implications for Research

These reflections on the defining character of human nature carry important implications for research with human animal chimeras. On the one hand, we gain a greater appreciation of both the continuity and inclusive breadth of human dignity. There is a compelling ethical imperative of protection of our distinctive nature across all stages of development and in every detail of our form and function; we are, through and through, beings of a rational nature. There is no single aptitude or attribute that defines us, rather, our species nature results from the full complement and synergistic powers of our distinctive human form. At the same time, we must acknowledge the crucial components of composition and complex dynamics at work in human development. From Systems Biology, we recognize that a precise relationship of material elements (*materia apta*) must be present for the integrated expression of the human form. This is not a matter of degree, we cannot be 99% human. As Aquinas observed, the form of a being signifies its essential presence: it is not like a fresco where the erasure of a part still leave half an ox.

This recognition of the unity, coherence, and contingent complexity of human nature sets a foundation and frame for considering the ethical and social significance of combining human and animal elements. Clearly, the first principle is that human embryos should never be considered suitable subjects for chimeric research – and no experiments with even the probability of creating an altered human embryo should be undertaken.

Where human elements are combined with animal embryos, we must give careful consideration to the likely outcome. Small genetic alterations, such as adding one or a few human genes, may in some cases rework basic aspects of form or physiology but are unlikely to express characteristics that challenge the boundaries of humanity. At the other extreme, most large-scale human–animal hybrids formed by mixing phylogenetically distant species will probably result in a failure of molecular coordination to a degree incompatible with the formation of a living organism. Recent studies of human–animal hybrids support such a conclusion. Stem cell biologist Robert Lanza reports that nuclear transfer of human nuclei into cow and rabbit eggs produced a pattern of gene expression dramatically distinct from natural embryogenesis and unable to support ongoing development (Chung et al. 2009). It is worth noting, however, that his studies did not include closely related species such as nonhuman primates. Here, the results might prove more ethically challenging.

Clearly, not all cellular growth need be regarded as embryogenesis. Mere cell division and even some evidence of progressive differentiation do not alone indicate the presence of an organism: germ cell tumors such as teratomas divide and differentiate. Even an enucleated egg will, when activated, sometimes proceed through several cell divisions. Like a spinning top, there is a certain molecular momentum carried within the maternally derived components of the egg. Likewise, limited signs of organization are not necessarily indicative of a full organism. Developmental biologist Roeland Nusse reports that embryoid bodies (self-organizing structures derived from embryonic stem cell cultures) show partial patterns of early gastrulation – though not the overall unity of a living organism (ten Berge et al. 2008). The artificial support of laboratory culture, like the womb itself, provides an environment in which nonorganismal life can be temporarily sustained – but, of course, in each of these situations the growing cells are completely dependent on one or more organisms (humans) for their protection and nurture.

In contrast, a natural organism is defined by an overarching integration of parts, a morphodynamic whole with a circular and self-referential network of relations. There is a purposeful cohesion, a global integration of action toward an end, a "top-down"[1] causation that supersedes and subsumes the parts into an irreducible and unified whole. An organism acts as the executive of its own existence; even within the dependence of the womb an embryo is rightfully considered a self-subsistent being. Its basic program of development is self-regulating and robust, if entrains the disparate parts within the defining form. For example, in an experimental technique labeled "whole embryo culture," when human mesenchymal stem cells are placed within the areas of the developing rat kidney during early development, the human cells are incorporated into the developmental program of the rat organism – as the authors report, the human cells "differentiate and contribute to functional complex structures of the new kidney" (Yokoo et al. 2005).

[1]While this concept of causation is often described as "hierarchical" or "top down," it is more rightly recognized as a "systems-causation," a seamless integration of action by the whole for the sake of the whole where causation is not exhaustively determined by the intrinsic properties of the particular parts.

This suggests that in any early chimerization, the dominant organismal form will order the basic architectural elements of the developing creature toward its species specific type. For this reason, early aggregations of human cells into predominantly animal embryos are unlikely to produce dramatically evident expressions of human form or function. Of course later fusion or transplantation of embryonic components (as with the geep or with the studies of Balaban and DeGoulian) will produce a recognizable patchwork of parts, but this is probably possible only with closely related species.

These reflections, taken together, suggest that there may be many scientifically useful and morally acceptable projects involving human–animal hybridization and chimerization. Clearly, each experiment will need to be evaluated ahead of time to assess the likely outcome and monitored for unanticipated results. Projects must be evaluated according to the seriousness of their purpose and the degree to which they involve the central defining characteristics of human life. Dimensions of form and function that carry a special human meaning (such as reproduction) or serve as signs that communicate the presence of a person (such as faces and voices) will raise the greatest concern. And, of course, there may be absolute prohibitions where human consciousness or other distinctive dimensions of human cognition are approached. With accumulated experience, we will clarify and refine our concepts of the boundaries of humanity and, no doubt, gain a greater appreciation for the distinctive dignity of the human form.

10.6 Policy Proposals

In light of the discussion above, the following principles are proposed for consideration as the foundation for public policy governing the mixing of human genes and cells in projects of cloning and stem cell research:

1. No production for research purposes of normal human embryos, or anything that might have the natural capacity to develop as an identifiable human organism. A criterion of nonviability cannot define a boundary of moral acceptability here because this would allow for the production of fetal monsters.
2. Great caution where there is ambiguity of organismal status or species identity – as with hybrid cloning – mere improbability of creating a human embryo is an inadequate criterion of justification.
3. No implantation of a human or modified human embryo into an animal womb. No implantation into an animal womb of any entity capable of developing with identifiable dimensions of defining human features of form or function – this goes not just for neurological capacities, but for body parts as well. No chimeric creatures should be produced that even rises to the level of having the basic human body plan, dominant human cell types, human anatomical infrastructure, or a majority of human organs.

4. No implantation into a human womb of any entity formed with human and animal genes, cells or tissues; nor of aggregations or combinations of human blastomeres or embryonic stem cells; nor with gametes exogenously produced from embryonic stem cells or genetically modified cells.
5. No implantation of animal embryos aggregated with human cleavage stage blastomeres (which could potentially contribute a human germ line).
6. Implantation of animal embryos with human embryonic stem cells injected into the blastocyst *only* when the human cells have been genetically altered to clearly restrict developmental potential, assuring no dominance of body plan and only limited lineages of human cells (and no human germ cells).
7. Later stage transplantation of human embryonic stem cell-derived tissues, cells, or organ primordia would be allowable until they manifest evocation of defining human dimensions of form or function. No chimeras with unique human neurological capacities – no human faces, larynx, hands, or genitals; no characteristic body plan, posture, or gait should be produced from such a project.
8. Late stage, post-anatomical transplant of animal parts into humans should only be done in order to heal or restore human form or function, never to alter, enhance, or degrade it – no animal sense organs, muscular or skeletal enhancements, no hooves, horns, feathers, fur, fangs, or antlers, and definitely no tails.
9. Where possible, in research involving the creation of chimeric creatures use phylogenetically distinct and distant species.
10. Err on the side of caution and preservation of natural boundaries, identifiable categories and kinds, thereby preserving our natural sense of an ordered and intelligible world.

10.7 Conclusion

Above all we must be concerned about our slow but steady drift toward treating all of living nature (including human nature) as mere matter and information to be reshuffled and reassigned for projects of the human will. While the prospects of human–animal mixing are troubling, they may offer extraordinary new avenues for progress in science and medicine as well as an opportunity to think more deeply about our human place in nature. In this way, they will provoke the reflection necessary for a more comprehensive definition and fuller appreciation of the meaning of being human – and a defense of human dignity.

References

Almeida-Porada G, Porada CD, Chamberlain J, Torabi A, Zanjani ED (2004) Formation of human hepatocytes by human hematopoietic stem cells in sheep. Blood 104(8):2582–2590

Anonymous (1984) It's a Geep. Time: http://www.time.com/time/magazine/article/0,9171,921546,921500.html

Ashley BM, DeBlois J, O'Rourke KD (2006) Health care ethics: a catholic theological analysis, 5th edn. Georgetown University Press, Washington DC
Brown DE (1991) Human universals. McGraw-Hill, New York
Chamberlain J, Yamagami T, Colletti E, Theise ND, Desai J, Frias A, Pixley J, Zanjani ED, Porada CD, Almeida-Porada G (2007) Efficient generation of human hepatocytes by the intrahepatic delivery of clonal human mesenchymal stem cells in fetal sheep. Hepatology 46 (6):1935–1945
Chung Y, Bishop CE, Treff NR, Walker SJ, Sandler VM, Becker S, Klimanskaya I, Wun WS, Dunn R, Hall RM, Su J, Lu SJ, Maserati M, Choi YH, Scott R, Atala A, Dittman R, Lanza R (2009) Reprogramming of human somatic cells using human and animal oocytes. Cloning Stem Cells 11(2):213–223
Congregation for the Doctrine of the Faith (2008) Instruction *Dignitas Personae* on Certain Bioethical Questions. http://www.vatican.va/roman_curia/congregations/cfaith/documents/rc_con_cfaith_doc_20 081208_dignitas-personae_en.html
Deacon TW (1990) Rethinking mammalian brain evolution. Integr Comp Biol 30(3):629–705
Dekel B, Amariglio N, Kaminski N, Schwartz A, Goshen E, Arditti FD, Tsarfaty I, Passwell JH, Reisner Y, Rechavi G (2002) Engraftment and differentiation of human metanephroi into functional mature nephrons after transplantation into mice is accompanied by a profile of gene expression similar to normal human kidney development. J Am Soc Nephrol 13(4):977–990
Dobrinski I, Avarbock MR, Brinster RL (2000) Germ cell transplantation from large domestic animals into mouse testes. Mol Reprod Dev 57(3):270–279
Emes RD, Pocklington AJ, Anderson CN, Bayes A, Collins MO, Vickers CA, Croning MD, Malik BR, Choudhary JS, Armstrong JD, Grant SG (2008) Evolutionary expansion and anatomical specialization of synapse proteome complexity. Nat Neurosci 11(7):799–806
Enard W, Khaitovich P, Klose J, Zöllner S, Heissig F, Giavalisco P, Nieselt-Struwe K, Muchmore E, Varki A, Ravid R, Doxiadis GM, Bontrop RE, Pääbo S (2002) Intra- and interspecific variation in primate gene expression patterns. Science 296(5566):340–343
Fleming N (2007) Go-ahead signalled for animal-human embryos. The Daily Telegraph, 1st March 2007, p 8.
Greely HT, Cho MK, Hogle LF, Satz DM (2007) Thinking about the human neuron mouse. Am J Bioeth 7(5):27–40
Hopkin M (2007) UK set to reverse stance on research with chimeras. Nat Med 13(8):890–891
Hurlbut WB (2005) Altered nuclear transfer. Stem Cell Rev 1(4):293–300
Joyce RE (1978) Personhood and the conception event. New Scholasticism 52(1):97–109
Karpowicz P, Cohen C, van der Kooy D (2005) Developing human-non human chimeras in human stem cell research: ethical issues and boundaries. Kennedy Inst Ethics J 15(2):107–134
Karpowicz P, Cohen CB, van der Kooy D (2004) It is ethical to transplant human stem cells into nonhuman embryos. Nat Med 10(4):331–335
Long KD, Kennedy G, Balaban E (2001) Transferring an inborn auditory perceptual predisposition with interspecies brain transplants. Proc Natl Acad Sci USA 98(10):5862–5867
Miller G (2005) Genetics. Mouse with human chromosome should boost Down syndrome research. Science 309(5743):1975
Molnár Z, Métin C, Stoykova A, Tarabykin V, Price DJ, Francis F, Meyer G, Dehay C, Kennedy H (2006) Comparative aspects of cerebral cortical development. Eur J Neurosci 23(4):921–934
Rose M (1997) Neandertal DNA. Archaeology:http://www.archaeology.org/online/news/dna.html.
Rowe ML, Goldin-Meadow S (2009) Differences in early gesture explain SES disparities in child vocabulary size at school entry. Science 323(5916):951–953
Savulescu J (2003) Human-animal transgenesis and chimeras might be an expression of our humanity. Am J Bioeth 3(3):22–25
Shreeve J (2005) The Other Stem-Cell Debate. The New York Times:http://www.nytimes.com/2005/2004/2010/magazine/2010CHIMERA.html.

Suzuki N, Sugawara M, Sugimoto M, Oshimura M, Furuichi Y (1997) Gene expressions of transferred human chromosome 8 in mouse cell lines. Biochem Biophys Res Commun 230 (2):315–319

Tamaki S, Eckert K, He D, Sutton R, Doshe M, Jain G, Tushinski R, Reitsma M, Harris B, Tsukamoto A, Gage F, Weissman I, Uchida N (2002) Engraftment of sorted/expanded human central nervous system stem cells from fetal brain. J Neurosci Res 69(6):976–986

ten Berge D, Koole W, Fuerer C, Fish M, Eroglu E, Nusse R (2008) Wnt signaling mediates self-organization and axis formation in embryoid bodies. Cell Stem Cell 3(5):508–518

Thomas K (1984) Man and the natural World: changing attitudes in England 1500–1800. Penguin Books, London

Travis J (1997) How fowl: chickens that behave like quail. Sci News 151(11):158

Weiss R (2005) U.S. denies patent for a too-human hybrid. 13th Feb 2005, p A03

Yokoo T, Ohashi T, Shen JS, Sakurai K, Miyazaki Y, Utsunomiya Y, Takahashi M, Terada Y, Eto Y, Kawamura T, Osumi N, Hosoya T (2005) Human mesenchymal stem cells in rodent whole-embryo culture are reprogrammed to contribute to kidney tissues. Proc Natl Acad Sci USA 102(9):3296–3300

Chapter 11
Is this Cell Entity a Human Being? Neural Activity, Spiritual Soul, and the Status of the Inner Cell Mass and Pluripotent Stem Cells

Antoine Suarez

Abstract Recent experiments demonstrate that somatic cells can, in principle, be reverted to a totipotent stage. This result stresses the need for rules allowing us to distinguish between embryos and nonembryos.

In this chapter, I discuss more in depth the assumption that the *proper biological potential* for developing neural activity specific to a human body's spontaneous movements provides the observable basis for ascertaining the presence of a spiritual soul. I show that this assumption does not mean reducing the spiritual soul to a material principle.

Additionally, I argue that the inner cell mass (ICM) of a human embryo shares the status of a personal human being; by contrast the trophectoderm (TE) can be replaced without changing the personal identity of the embryo (just as changing the heart or skin of an adult does not change his or her personal identity). I show however that deciding this argument requires new experiments in animal models. These experiments are crucial for distinguishing between embryos and nonembryos.

Finally, I stress that for coherently deciding whether a cell entity shares the status of a human person it is crucial to take the human body's behavior as the basis for philosophical reflection and ascription of rights.

Keywords Human spiritual soul • Induced pluripotent stem cells (iPSCs) • Inner cell mass (ICM) • Neural activity • Pluripotency • Proper biological potential • Totipotency

A. Suarez
The Institute for Interdisciplinary Studies, Berninastr. 85, 8057 Zurich, Switzerland

Social Trends Institute/Bioethics, Barcelona, Spain

Swiss Society of Bioethics, Zurich, Switzerland
e-mail: suarez@leman.ch

A. Suarez and J. Huarte (eds.), *Is this Cell a Human Being?*,
DOI 10.1007/978-3-642-20772-3_11,

11.1 Introduction

In the previous chapter 5, "Embryos grown in culture deserve the same moral status as embryos after implantation" (Huarte and Suarez 2011) I concluded, along with Joachim Huarte, that:

- The "proper biological potential for developing neural activity specific to a human body" provides the observable basis for ascertaining the presence of a spiritual soul.
- The presence of DIANA insufficiencies in a cellular entity's genomic information (insufficiencies that Directly Inhibit the Appearance of Neural Activity) ought to be considered a sure sign that such a cellular entity is not spiritually ensouled and, therefore, is not a human being.

The aim of this particular chapter is to discuss, in greater depth, the philosophical basis of these conclusions. Hence, I will argue that recent achievements with induced pluripotent stem cells (iPSCs) reinforce the relevance of the DIANA criteria (Suarez et al. 2007). The generation of mice that were fully derived from induced pluripotent (iPS) cells demonstrates the possibility, in principle, of reverting adult somatic cells to a totipotent stage, and hence provides greater weight to the claim that we need some kind of criteria for distinguishing between embryos and nonembryos. Later on in this chapter, I will argue that discerning whether a particular reprogramming technique produces a human totipotent cell entity, and therefore an embryo, requires complementation experiments that combine cells (iPSCs) with trophectoderm (blastocyst without inner cell mass [ICM]) in animal models. Such experiments can be done using well-known techniques. I conclude by stressing the importance, to any criteria established so as to distinguish between embryos and nonembryos, of the following philosophical principle: that the human body's spontaneous behavior is the basis for deriving the categories that allow us to understand the world and organize society. Moreover, this spontaneous movements' metaphysics is the basis for a consistent metaphysics of the human body–soul relation.

11.2 The Neural Activity Responsible for Spontaneous Movements of the Body Cannot be Explained Exclusively by Material Agency

As was argued in our earlier chapter 5, the neural activity responsible for the spontaneous movements of a human body is a clear sign for ascertaining the presence of a human spiritual soul and thus plays a key role in our proposal for establishing that a cell entity or organism is a human being and a person. Against this principle, one might object that the neural activity in question is derived from the physiology of the brain and as such is something material and so it would be odd to identify the human soul with such an activity. In response to such an objection, I want to stress that although the neural activity responsible for the spontaneous movements of the human body is an observable effect (something we can see *in* the brain), its presence cannot be explained *solely* by the assertion that it was *caused* by pure material deterministic agency. Rather, as I want to argue, these effects must also be understood as the result of quantum physical agency that comes from outside of space–time. To defend this claim, it is necessary that we take a little detour through contemporary findings in quantum physics.

Recent studies tell us that quantum randomness is inseparable from nonmaterial principles acting and controlling randomness from outside space–time. In so-called entanglement experiments, correlated events appear in regions separated by a great distance. The distance is such that no material connection can explain the correlations. Things can best be described as follows: in a lab (say in Zürich), a physicist has a device for detecting photons (light particles) with two detectors denoted D0 and D1. For each photon, only one of the two detectors light up, that is, the measurement will read as yielding the result 0 (D0) or 1 (D1). After many test runs, the results obtained (1,0,0,1,0,. . .) are distributed in a way not dissimilar to the results one would get by tossing a coin (and recording the results as 1 for head and 0 for tails). In other words, they do not exhibit any particular pattern or order. In another lab (say New York) another physicist with a similar device gets similar results: 0 and 1 distributed in much the same way as one would get if they were just tossing a coin. So, there are random results in Zürich and random results in New York. Now, the results in Zürich and New York occur pair-wise: for each result in Zürich there is a corresponding result in New York occurring almost simultaneously. Because of the distance and insignificant time interval between the detection in Zürich and in New York, the two results could not be connected by any signal traveling at the speed of light. Nonetheless, when the physicists come together and compare their results they see that there is a perfect correlation between the two experiments: when the result in New York is 0, the result in Zürich is 0, and when the result in New York is 1, the result in Zürich is 1. Randomness in Zürich, randomness in New York, but, interestingly, the same randomness in both places!

In 1964, John Bell, by way of a magnificent mathematical result, found that it is possible to exclude that the photons behave like genetic twins determined by some “hidden genetic program” in order to explain the correlations (Bell 1987). This happens by way of more sophisticated experiments, in which each experimenter has the choice of switching his device into one of two different configurations. Depending on the experimenters’ choices, the results in New York and Zürich are correlated, but they are not correlated perfectly. Bell’s theorem imposes a limit to the degree of correlation hidden programs could achieve in this type of experiment. The degree of correlation experimentally observed and predicted by quantum theory violates Bell’s limit. This means that one cannot account for the correlations by means of common causes in the past, unless each experimenter himself is predetermined in choosing the settings of his apparatus. Hence, if one assumes that the experimenters are free, the correlations cannot be explained either by a common cause in the past or by any material signal traveling in space–time from one place to the other.

Even more sophisticated experiments involving devices in motion (relativistic experiments) demonstrate that the quantum correlations are independent of any temporal order (before–before experiment), and in this sense come from outside space–time (Stefanov et al. 2002, 2003; Branciard et al. 2008; Suarez 2008).

It is important here to note that one cannot account for such effects invoking "indeterminism" or "randomness" alone. Here, randomness and control appear to be inseparably united in the same phenomenon. That is, there is randomness in Zürich and randomness in New York, but it appears to be *steered* by some agency whereby "the same randomness" is produced in separate places. If I accept that I am free, then quantum experiments demonstrate effects in which the control of quantum randomness appears to happen from outside space–time, and in this sense it can also be considered an experimental proof on behalf of nonmaterial agency on the part of nature.

Of course, quantum physics does not *demonstrate* that we have "free will" or "consciousness." Anyone can decide this question for himself as he wishes. I may prefer to think that free will is an illusion and the movements of my fingers are the result of a chain of material causes reaching all the way back to the Big Bang. However, such a choice is not compulsory within the framework of quantum physics. Within this framework, nothing speaks absolutely against the assumption that the neural activity responsible for the chapter I am writing now cannot be explained exclusively by material or observable causal chains and so requires spiritual (nonmaterial) agency that occurs from outside space–time.

What follows from what was said above is that my position would hold that "freedom," when understood as being grounded in an immaterial principle (like the soul), is in principle compatible with the findings of contemporary quantum physics. Metaphysics is based on observations, and today's physics provides experiments that may inspire metaphysical reflection.

I may then ask myself another metaphysical question. Suppose I was to make the existential choice of believing that my freedom is not an illusion and therefore that my corporal movements reveal spiritual agency: how would I conclude that the body of someone else is also animated by a spiritual soul? When I fix my eyes upon a colleague with whom I am speaking, I conclude that the human being sitting in front of me is animated by a spiritual soul and is a person because (a) it has the same specific form (or shape) as my body, and (b) this form or shape exhibits movements similar to the movements I make for expressing thoughts, emotions, and even rights-claims. These movements are what Joachim Huarte and I have called "spontaneous movements." It is irrelevant whether such movements are conscious or not because the distinction between conscious and unconscious can be notoriously tricky, especially in regards to the movements performed by others.

Indeed, my own understanding of the actions of others follows from the concepts, emotions, etc., that I associate with such actions when I perform them myself. The neurophysiologic correlate of this philosophical supposition is the mirror neurons system, which unifies action perception and action execution:

"Each time an individual observes another individual performing an action, a set of neurons that encode that action is activated in the observer's own cortical motor system." (Rizzolatti and Sinigaglia 2010).

11.3 The Criterion of Spontaneous Movements Overcomes Dualism and Reinforces the Unity of the Human Being

Following what was said above, I want to stress that the emphasis on spontaneous movements suggests that we should *deny* that a cellular entity or organism is a human being if that cellular entity or organism has an intrinsic *incapability* for spontaneous motility. Such *incapability*, I argue, renders the entity incapable of animation by a human soul. More specifically, we can say that a person dies when the human body irreversibly loses the capability for the neural activity responsible for spontaneous movement.

I now want to anticipate an important objection that runs as follows: since the human being is a unity and the rational soul animates the entire body and to the extent that the biological platform for spontaneous movements is only a *part* of the entire body, not the entire body, why should its mere absence alone be sufficient to render *the whole of which it is a part* incapable of rational animation? Does this not amount to a position whereby one holds that the biological platform for spontaneous movements is, in fact, the organ of ensoulment, through which the rational soul only indirectly animates the rest of the human body? Such an ontology would imply the dissolution of the unity of human bodily architecture (of which the platform for spontaneous movement is only a part, albeit a very important part) and of the unity of the human being tout court. Such an ontology, in fact, appears to amount to a denial that it is the one human being who is the sole *suppositum* of all of human vital operations, regardless of whether or not they can directly manifest the rational soul qua rational.

To such an objection, I would like to offer the following response:

By stating that spontaneous movements reveal the presence of the soul, whereas for instance heart beating alone doesn't, we don't dissolve the unity of the human being, but simply stress that *this unity does not imply that changing the heart means changing the soul as well.*

The capability of spontaneous motion is not "only a part of the body" but a capacity without which there simply is no body. If there is capability of spontaneous movements, the heart beating is under steering of the human spiritual soul as well. If there is no capability of spontaneous movements, neither the heart's beating nor any organ's activity is under steering of the spiritual soul. By stressing the capability for spontaneous motion, I am not suggesting that the lack of capability for spontaneous movements renders the whole human body incapable of rational animation. We rather state that such a privation excludes, in principle, the possibility of having a human body at all inasmuch as it precludes any control by the immaterial soul.

In other words, if the capability for spontaneous motion is present, then there is one embodied human being who is the sole *suppositum* of all vital operations, even when these operations do not directly manifest rationality or consciousness. Patients in persistent vegetative state and children with anencephaly exhibit spontaneous movements even though they remain unconscious. According to the metaphysical view I laid out above, such patients are human beings and persons. If the capability for spontaneous movements disappears, then we no longer have one human being or embodied person and the remaining vital operations are no longer those of a human body but rather are the operations of an organism made of human cells that lacks the features specific to animal life and thus is not animated by a human soul. An organism to be a human being has to be an animal. If it displays "vegetal life functions only" it cannot be considered a human being just insofar as they lack the features specific to animal functioning. The body with a capability for spontaneous motility (the "living human body") is not just a "biological platform" for rational ensoulment, but rather is already the rationally ensouled human person herself. Neither the soul nor the body preexists the other; both come into being simultaneously. The body with human architecture and spontaneous motility is nothing other than the embodied presence of the soul in space and time.

The spontaneous movements should not be conceived as a reality separate from the soul, that is, as occurring randomly when not being acted upon by the soul and in an ordered way when the soul controls them. Such a conception is *dualistic*. Spontaneous movements are always an activity of the soul, even when they happen "at random." In other words, the randomness itself is generated by the soul. Just as quantum randomness is inseparably united to order/control originating from non-material agency, the neural activity responsible for spontaneous motility is inseparably united to the human spiritual soul (the controlling principle), even when the subject is not aware that he is acting. Spontaneous movements are actually ruled by the will even if they do not always originate in a conscious way and are not always immediately guided by some intention.

Let me offer an example of what I am trying to get at above. When little Suzy plays the piano, the spiritual powers of her soul are not responsible for generating the energy necessary to trigger the movements of her fingers, but are instead responsible for the order in which her fingers move. We could represent this order in digital form by a very large sequence of bits (1,0,0,1,1,0,1,0,0,. . .). This number-string could represent the digital transcription of the motor activity necessary to play Beethoven's Fifth. Now, it is of course often the case that this order is more or less deliberate, but one could also say that the "muscle memory" that develops from such deliberate playing allows little Suzy to play Beethoven in an almost unconscious way. Still, Suzy may unconsciously move her fingers as if she were playing the piano even though she is not currently sitting at her Steinway. Now, this sequence of movements also contains information corresponding to a sequence of bits (0,1,0,0,1,0,1,1,. . .), which is different than that which corresponds to the Beethoven Symphony. What I want to say is that the order of the numbers in both number-strings originates from the spiritual powers of Suzy's soul.

Similarly, when I whistle a song I am to a certain extent consciously controlling my breathing. Again, the particular sequence or order of these movements (not the energy needed to trigger and sustain them) comes from the spiritual powers of my soul. When I am sleeping, the sequence or order of my respiratory movements, even if they have no particular meaning, comes from my spiritual powers even though it is accompanied by a very low level of consciousness. Of course, there are many possible sequences of breathing and lip movements that can produce no meaning at all. But the fact that *this* particular sequence occurs instead of another depends on the will. I regulate the movement of my lips through my free will when I move them consciously as well as when I move them unconsciously. In this sense, random spontaneous movements of a human body are nothing other than a particular expression of human will: they are unconscious voluntary movements. This conclusion is in accord with Benjamin Libet's experiments, which demonstrate that voluntary acts are not necessarily conscious ones (Libet 2002).

According to Thomas Aquinas, babies and irrational animals participate in the "voluntary" though in an imperfect way (STh I-II 6.2). I think that unconscious spontaneous movements can be considered to be "imperfect voluntary movements" as well. So, one can assume that the spontaneous motility a fetus exhibits in the 7th week of pregnancy (De Vries et al. 1982) is somewhat voluntary (Suarez 1993). I must stress that this does not mean that the will as a spiritual potency triggers such movements. Consider for instance the saccadic movements of the eyes. They are paramount indicator of the presence of spontaneous motility and they are the movements most directly influenced by the brain. The irreversible disappearance of these movements is a main criterion for establishing death. The activation of such movements (the energy required for performing them) does not come from the will. The will, as a spiritual power, only influences the particular sequence (order) in which such movements happen. The view that spontaneous movements are voluntary ones, happening in a more or less conscious way, seems to be supported by Aquinas' idea of "imperfect voluntary movements." A consequence of admitting that spontaneous movements (although imperfectly) are always somewhat commanded by the will is that they reveal nonmaterial (spiritual) agency.

But then why does Thomas Aquinas argue in favor of the hypothesis called "delayed hominization" and accepts the possibility of a middle animal stage in which the human embryo was without an "intellectual soul"? I think the reason is that Aquinas (STh I, 76.3 and 118.2) following Aristotle (GA) admits the possibility of an animal organism made of human flesh (organic matter), which is not a human being, a body animated by a human spiritual soul. "Delayed hominization" is surely a consequence of incomplete biological knowledge (Pangallo 2007), but probably mainly comes from assuming that in the generation of man there can be an animal which is not a man, i.e., an animal without human spiritual soul.

It is a matter of philosophical perspective: Aquinas and Aristotle seem to explain the human body and its generation on the basis of categories they take from the other animals, while in this chapter we try rather to characterize the animal behavior on the basis of categories derived from the *human body's spontaneous behavior* and its relevance for defining coherently the personal rights. Our basic category is not "animal life" but that of "spontaneous movements of the human body": We then define animals as living organisms exhibiting movements that look like the human spontaneous ones. If one accepts that any human animal organism is animated by a spiritual soul, then one has to conclude that a human embryo capable of spontaneous movements is animated by a human spiritual soul (see Sect. 11.4 below).

By contrast, one cannot avoid delayed hominization as long as one assumes that animal features of the human body like sense and spontaneous motility do not involve any use of the powers of the soul. Indeed, by this assumption one is even led to admit the possibility, in principle, of an organism made of human cells and sharing features specific to animal life, which is not animated by a spiritual soul. Such a position is at risk of being used today to argue that hydranencephalic children or adults, patients in permanent vegetative state, and even fetuses and babies (as long as they cannot perform intellectual operations) are not spiritually ensouled and therefore lack any significant moral status. (For a characterization of the conditions of hydranencephalic children or adults and patients in permanent vegetative state, see Counter 2007 and Laureys 2005).

Now, suppose that in order to avoid this problem one adopted the following *position*: Though only intellectual capabilities (like speech) reveal spiritual powers, spontaneous motility (breathing, movements of eyes and legs, etc.) indicates that the organism is alive, and therefore, "because of the uniqueness of the soul as principle of life," one has to conclude that the organism's soul is also the spiritual one. However, this *position* itself is inconsistent.

Indeed, Thomas Aquinas does not think anything in the hypothesis known as "delayed hominization" contradicted the supposition that the soul qua form of the body was utterly unique and one (that is, the hylemorphic substantial unity of the human composite). In the end of the hominization process there is *only one* soul, which is at once vegetative, sensitive and intellectual, while in the middle animal stage there is only one soul as well, which is at once vegetative and sensitive. Thus, Thomas' conclusion (that there can be organisms of human origin that exhibit animal life and are not spiritually animated) cannot be circumvented by invoking the uniqueness of the soul as principle of life. To avoid "delayed hominization" one needs something more, that is, to assume that the characteristics of human animal life reveal by themselves animation by a spiritual soul. One could alternatively assume that functions characteristic of human vegetal life (growth, heart beating, circulation, reproduction) reveal by themselves the presence of a soul with spiritual powers. However, this implies that transplanting the heart of an individual means transplanting his spiritual soul, and leads additionally to the rejection of the current clinical criteria for defining death.

The argument makes clear the importance of what characteristics of human life one chooses as the observable basis for predicating personhood *directly* to a body. On my view, one should not predicate personhood of a body (of human origin) on the basis of intellectual capabilities (like speech) nor should one predicate personhood on the basis of vegetal life functions *alone*, but rather one should predicate personhood of a body (of human origin) on the basis of the specifically animal feature of bodily life – spontaneous motility. The assumption that spontaneous movements directly reveal a body's animation by a spiritual soul:

- Contains by itself the uniqueness of human soul, as stated above: If the capability for spontaneous motion is present, then there is one human being who is the subject of all vital operations.
- Excludes the possibility of an organism made of human cells that shares the features of animal life and is not spiritually animated.
- Does not exclude the possibility of an organism made of human cells that shares exclusively features of vegetal life and is not spiritual animated.

In summary, it seems to me that dissolving the unity of spontaneous movements and the spiritual powers of the soul is not profitable for deciding whether a cell entity or organism of human origin is a human being properly speaking or not, and therefore is not of great benefit for settling arguments related to the beginning and the end of the temporal existence of the human person.

11.4 The Principle of *Conservation of Personal Identity* During Growth Allows Us to Move Consistently from Spontaneous Movements to Capability for Them

In the previous sections, I argued that the presence of spontaneous motility in a cellular entity of human origin is *per se* revelatory of its animation by a spiritual soul. Is it consistent, however, to conclude from this claim that the *proper biological potential* for spontaneous motility is already a sufficient biological condition for establishing spiritual ensoulment? In this section, I argue that such a move is appropriate and is grounded in the principle of *conservation of personal identity* during bodily growth.

Anyone claiming to have a personal identity implicitly accepts that his or her identity is conserved in time. As an author of this chapter, for example, I claim to be the same person as the author of other papers I wrote years ago. That is, my personal identity has roots outside time. In this sense I, and the paper I am writing now, cannot be explained exclusively by material or observable causal chains (see Sect. 11.2). If someone admits only material causal chains within time, he will deny that personal identity persists through time.

When we establish that a child B and an adult B' are the same person we assume that bodily growth does not amount to the appearance of a new person: for example, if the body B grows into B', and B' is a person, than B shares the same personal identity as B'. In more scientific terms: if the cells of B' originate from the cells of B, B and B' are the same person. The assumption of a spiritual soul is coherent with the principle of conservation of the personal identity.

However, today it is also generally admitted that a human body B which receives a foreign heart does not undergo a change in his or her personal identity (even if the person can be affected by profound psychological changes in his/her feelings, tastes, preferences, etc.). The fact that the heart's cells of B' (with the new heart) do not originate from the heart's cells of B does not prevent conservation of personal identity. The same holds for transplantation of other organs. This view is based on the implicit assumption that the brain centers responsible for the spontaneous motility of B' derive from the brain centers responsible for the spontaneous motility of B.

In a more precise way, I want to argue that one can establish *conservation of personal identity* during bodily growth as follows:

Consider that a cell entity or cell layer X develops into a body B exhibiting human architecture and spontaneous motility, where this development requires the interaction of X with other cell clusters or cell layers in its environment. I would argue that X *shares the same personal identity* (is animated by the same spiritual soul) as B if the following two conditions are met:

1. The neural centers of B responsible for its spontaneous activity derive exclusively from cells contained in X.
2. Equivalent neural centers cannot be derived from the cell clusters or cell layers interacting with X.

As regards the cell clusters or cell layers interacting with X, I want to suggest that they share a similar organic status as that of the heart or skin of B, that is, they can be replaced without loss of personal identity. In other words, I define that X has the *proper biological potential* to develop into B *if* the observable biological features (like genetic and epigenetic information) allow us to predict that conditions (1) and (2) above will be met. In other words, that "X has the *proper biological potential* to develop into B" is just another way of saying that "X and B share the *same personal identity*." We also say, following these principles, that X is *totipotent*.

The sense of the term *proper biological potential* has to do with an active potency that is rooted in the substantial form of the entity in question. This active potency is revealed by a certain material organization but is not ontologically identical with this material organization. Proper biological potential as an active potency, rooted in substantial form, is not spatio-temporal, though of course it needs a particular kind of material organization for its unfolding in space and time. Just as breathing reveals that a spiritual human soul is controlling the breathing observed, the appropriate genomic (genetic and epigenetic) information reveals that the biological potential (and the corresponding spiritual soul) is there.

So the *proper biological potential* for human spontaneous motility is more than an adequate bodily condition – a "dispositio corporis" (Pangallo 2007) – for spiritual ensoulment, it is a visible sign revealing the presence of the invisible spiritual soul. A human cell entity sharing such a potential is more than appropriate matter – "materia apta" (Walker 2005) – for receiving the human spiritual soul, it is already the presence of this soul in space and time, a human animal animated by a spiritual soul, an embodied person.

At this point, it is worth taking note of two alternative criteria for establishing personhood: the capability for conscious communication and capability for blood circulation in a human body. As to the former, one would need to go back and establish what counts as the proper biological potential for speaking as the appropriate condition for ascertaining animation by a spiritual soul. Accordingly, embryos of hydranencephalic children would not be persons. As said in Sect. 11.3, this conclusion seems unacceptable to us. Similarly if one takes heart beating and blood circulation as the observable feature of the human body that directly reveals the spiritual soul, then one would have to conclude that the proper biological potential for heart beating is the appropriate condition for ascertaining spiritual ensoulment. Accordingly, parthenotes (see Sect. 11.6) and brain dead organisms should be considered persons. As I shall argue in the Sect. 11.6, I think this is a conclusion we should reject as well.

11.5 On the Status of the ICM and the Trophectoderm (TE)

To illustrate how the principles enunciated in the preceding Sect. 11.4 might play out as applied to a particular case, I want to ask how one might establish the status of the ICM and the status of trophectoderm cell layer (TE) in the developing blastocyst (the embryo before implantation into the maternal uterus).

The blastocyst consists of two distinct cell layers: the ICM, which becomes the embryo proper and the adult human being; and the TE, which contributes to the placental support system. Although the TE is essential for the further development of the embryo, it does not become part of the full-grown organism. Consider now a body B exhibiting human architecture and spontaneous motility that derives from a normal developing blastocyst. Before implantation into the uterus, the ICM and the TE interact with each other. After implantation, the uterus interacts with the TE and through it with the ICM as well. The neural centers controlling the spontaneous movements of the body B derive exclusively from the ICM through cell division. No body exhibiting human architecture and spontaneous motility can derive from an implanted TE alone through cell division. If one separates the ICM from the TE, the ICM gives rise to a colony of embryonic stem cells (ESCs) and the TE degenerates into a tumor. From observations in mice one knows that the (fertilization-derived) ICM in combination with a different healthy TE' produces a blastocyst ICM + TE' developing into a healthy adult B'.[1] So, as these studies show, the neural centers controlling the spontaneous movements of the body B' derive exclusively from the ICM cell layer.

[1] See the control experiments in Gardner et al. (1990). Actually, in these experiments one combined the ICM of a blastocyst with TE and PE (Primitive Endoderm) of another blastocyst. However, since PE like TE gives rise to only extraembryonic tissue, this particular does not change our conclusion and for the sake of simplicity we dismiss it in the argument. Thus, hereafter in this chapter the notation 'TE' means 'TE+PE' (trophectoderm and primitive endoderm), unless stated otherwise.

Such observations prove that if B and B' exhibit spontaneous motility, and not only autonomous and reflex activities (like heart beating, blood circulation, kidney activity), it is because the neural centers derive from the cell division specific to the ICM. This means that the ICM itself has the *proper biological potential* and is the *key* determinant for the identity of the whole embryo. I think that this point has two important implications: *first,* the ICM without a TE is the same human being (has the same personal identity) as the body B, even if the ICM without a TE will die very quickly much in the same way that the embryo without a functioning uterus is the same human being as B, even if the embryo without a functioning uterus will die quickly; *second*, B' is the same human being (has the same personal identity) as B and the ICM. The trophectoderm TE can be replaced without changing the personal identity of the cellular organism, just as changing the heart or skin of B does not change his or her personal identity.

If one were to hold a position that the ICM does not determine the personal identity of the whole embryo, it would produce quite a few oddities. Such a position would suppose that an ICM + TE and an ICM + TE' do not have the same personal identity, that they are two different persons. If one holds this position, then either one assumes that the identity of a premature infant changes after one removes the placenta and introduces him or her into a neonatal intensive care unit, or one has the burden of explaining at what moment of development (and why) the placenta could be removed without changing the personal identity of the fetus. On the view I sketched above, it would be my own opinion that the TE (and the placenta that is eventually generated by the TE) has no bearing whatsoever on the personal identity of the developing embryo, fetus, or newborn.

11.6 On the Status of Cell Entities Obtained Through Altered Nuclear Transfer and Parthenotes

Consider the following scenario: a process of fertilization (or an artificial replication thereof, parasitic on normal reproduction) begins, but is prevented from realizing its teleology, that is, the end to which that process is directed. The result of this failure is a cell entity X. How can we be sure that, in the interval of time between the beginning of the fertilization and the actual failure of what would have been its normal outcome, X was never a human being, if only for a split second? Such a question implicitly assumes that the process of fertilization (or the artificial replication thereof) produces a cell entity X* that is a human being. However, X* is then immediately prevented from realizing its developmental potential and the result of this failure is X. In cell entity X* it is always possible to distinguish between two cell lineages: one leading to the ICM* (from which the neural centers derive) and the other to the TE* (from which only placental tissue derives). According to the principle of *conservation of personal identity* (see Sect. 11.4), assuming that X* is a human being is the same as assuming that the lineage leading to ICM* is healthy. Accordingly, one must assume, if indeed X was at one time X*, that X* was prevented from realizing its developmental potential due to a defective TE* lineage. If one combines the ICM* lineage of X* with a healthy lineage TE°, the combination ICM* + TE° will normally develop to birth.

This line of reasoning leads us to three distinct conclusions: first, the human being deriving from ICM* + TE° is the same person as X*; second, X has the status of a brain dead organism resulting from the death of X*; third, the cause preventing X* from realizing its teleology is not a DIANA defect. In other words, if cellular entity X (that is not a human being) originates from a former cell entity X* which was a human being (who then died), then the cause preventing X* from realizing its teleology is *not* a DIANA defect, i.e., one that Directly Inhibits the Appearance of Neural Activity.

Additionally, the preceding reasoning leads to the following objection against altered nuclear transfer (ANT) through inactivation of the *Cdx2* gene (Meissner and Jaenisch 2006):

Consider the first cell division: One of the cells (with expression of the *Cdx2* gene) leads to the TE lineage, while the other (with expression of the *Oct4* gene) leads to the ICM lineage.

If before fertilization one prevents expression of *Cdx2*, one produces a blastocyst with a defective TE and healthy ICM. Assume that if you combine this ICM with a healthy TE (the experiment has not yet been done), you get development to birth. If so, it would mean that inactivation of *Cdx2* is not a DIANA defect, that is even if one prevents expression of *Cdx2* before fertilization, it is still the case that the produced ICM is a human being.

If before fertilization one deletes *Oct4*, this yields a blastocyst without an ICM (and therefore not useful for deriving ESCs), which gives rise to a complete mole consisting *only* of extra-embryonic tissue. Deletion of *Oct4* is a DIANA defect. I conclude that deleting *Oct4* before fertilization leads after fertilization to a blastocyst that is not a human being.

Consider now the case of *standard parthenotes* (with two nuclei of fully grown oocytes) which develop into a blastocyst with a defective TE and a well-developed ICM (Huarte and Suarez 2011). After implantation, they degenerate and die. But that they died is not enough of a basis on which to conclude that standard parthenotes are not human beings. Only if I combine the ICM with a healthy TE*, and observe that the reconstituted blastocyst ICM + TE* develops to a heart beating stage but then stops can I conclude that the standard parthenotes are not human beings. Experiments in animal models show that this is the case (Gardner et al. 1990). *Standard parthenotes* are incapable of developing spontaneous motility and can be considered to share the status of a brain dead human organism. I think this conclusion is further supported by the case reports of teratomas that resemble a malformed fetus and exhibit a highly differentiated homunculus of parthenogenetic origin (Kuno et al. 2004; Weiss et al. 2006). Fetiform ovarian teratomas are thought to arise from parthenogenetic development of a primordial germ cell. In this sense they may demonstrate that a germ cell is capable of producing any tissue found in the human body (Weiss et al. 2006), but do not prove that a germ cell or an oocyte have the proper biological potential of developing into a living fetus.

Standard parthenotes should not be confused with the so-called *ng-parthenotes* (Suarez et al. 2007; Huarte and Suarez 2011). These are produced by altering the nucleus of a nongrowing oocyte and then transferring it into a fully-grown oocyte (Kono et al. 2002, 2004). Such a nucleus shares decisive characteristics with the nucleus of a male gamete, and its transfer into a fully grown oocyte is very much equivalent to fertilizing an egg. The resulting artifacts reach spontaneous motility and even birth. According to our DIANA criterion, *ng-parthenotes* are disabled embryos and in humans would share the same moral status as children with anencephaly.

In summary, *standard parthenotes* deriving from a human egg can be considered a "human organism" but not a "human animal," and therefore one can conclude that they are not human beings. In contrast human *ng-parthenotes* and ANT cells obtained through inactivation of the Cdx2 gene would be "human animals" (animals with a human spiritual soul) and deserve the moral status of persons.

11.7 When Does Reprogramming Somatic Cells Produce a Human Being?

Criteria for distinguishing between embryos and nonembryos were first motivated by alternative proposals to derive embryonic-like pluripotent stem cells without destroying human embryos. In this context it was important to ascertain whether the proposed techniques produced a disabled human being or a cell entity that is not a human being.

The arrival of induced pluripotent stem cells (iPSCs) obtained through reprogramming of adult somatic cells was wholeheartedly cheered as a means of producing embryonic-like pluripotent stem cells without the necessity of destroying embryos. One could think that with the arrival of iPSCs on the scene, distinguishing between embryos and nonembryos would remain only a question of academic interest. However, work underway with iPSCs is showing a greater interest in answering this question and so researchers working in that area may also appreciate attempts (such as DIANA) to provide scientists, ethicists, and policymakers with a normative criterion for distinguishing between embryos and nonembryos.

Combining embryonic stem cells (ESCs) with a tetraploid blastocyst (tetraploid complementation) (Rossant and Spence 1998) or laser-assisted injection of ESCs into an eight cell-stage diploid embryo (Poueymirou et al. 2007) makes it possible to generate mice fully derived from the ESCs. Experiments show that it is possible to use tetraploid complementation to generate mice that are fully derived from iPS cells (Zhao et al. 2009; Boland et al. 2009; Kang et al. 2009) as well as from parthenogenetic embryonic stem cells (pESCs) (Chen et al. 2009).

These results mean that it is possible, in principle, to go beyond pluripotency and reprogram adult cells to a totipotent stage as well (Brugger 2010). But then the question arises: how can we be sure that the iPS cells resulting from a particular reprogramming technique are not a human being? I want to now apply some of the principles and arguments established above in order to offer a possible answer to this question.

A tetraploid blastocyst consists of a tetraploid inner cell mass (tICM) and a tetraploid trophectoderm (tTE). On the one hand, tetraploid human embryos can implant and develop to an adult whose cells are fully derived from the tetraploid ICM (Nakamura et al. 2003, and references therein). Additionally, in tetraploid complementation as typically performed in mice, the tetraploid ICM is not removed from the blastocyst before the injection of the ESCs, iPSCs, or pESCs (Zhao et al. 2009; Kang et al. 2009; Chen et al. 2009). This means that on the basis of the available observations one cannot exclude that ESCs, pESCs, or iPSCs require the interaction with fertilization-derived ICM cells *additionally* to TE cells in order to become an adult.

On the other hand, experiments performed in 1990 by Gardner and others using reconstitution methods show that combining pESCs with fertilization-derived trophectoderm (TE) (i.e., a blastocyst from which the ICM has been removed) develop, at best, up to the stage of heart beating and yolk sac circulation, but are not capable of going beyond that stage (Gardner et al. 1990). In particular, such reconstituted blastocysts pESCs + TE fail to reach spontaneous fetal motility.

These results with parthenotes show that the parthenogenetic ICM lacks totipotency. The fact that parthenogenetic ICM cells possess a similar type of pluripotency to embryonic stem cells (Brugger 2010) does not allow us to conclude that *standard parthenotes* are embryos.

Additionally, these results strongly support the hypothesis that the fertilization-derived ICM is crucial in order to make stem cells totipotent, i.e., capable of generating a living organism exhibiting spontaneous motility.

One could test this hypothesis by means of an experiment in animal models using reconstituted blastocysts ESCs + tTE (combination of embryonic stem cells with tTE) and iPSCs + tTE (combination of induced pluripotent stem cells with tTE), which can be obtained through reconstitution techniques similar to those used by Gardner and co-workers in 1990. As control experiments, one could combine fertilization-derived ICM with tTE, i.e., blastocysts ICM + tTE. We remember that by 'TE' we are denoting 'TE+PE' (see footnote 1 earlier), and therefore the experiment we are proposing involves actually triple reconstituted blastocysts (ESCs + tTE + tPE; iPSCs + tTE + tPE; ICM + tTE + tPE) like in the experiment by Gardner et al. 1990.

I predict that using the ESCs and iPSCs available today, reconstituted blastocysts with ESCs + tTE and iPSCs + tTE will fail to develop (in more or less the same way that blastocysts reconstituted with normal TE, PE and parthenogenetic ICM fail to develop in the Gardner experiments), while ICM + tTE will develop to term.

Unpublished results by Richard Gardner and Janet Rossant have shown that simply injecting clumps of ES cells inside vesicles of pure trophectoderm (i.e., a blastocyst from which the ICM and the PE have been removed) leads to cell entities that fail to develop (private communication of February 10th, 2010, respectively

June 7th, 2011). These results speak in favor of my prediction. Nonetheless one could object that it is the primitive endoderm (PE) that helps the ESCs to survive and make a fetus. To exclude this possibility it is necessary to perform an experiment using triple reconstituted blastocysts (ESCs + tTE + tPE; iPSCs + tTE + tPE; ICM + tTE + tPE) as well.

In any case, if my prediction holds, then human ESCs and iPSCs cannot be considered to be totipotent on their own, i.e., they do not have the *proper biological potential* to develop into a human baby. By contrast, the ICM layer should be considered *totipotent* and sharing the *proper biological potential* to develop into a human fetus, newborn, child, adult and is therefore revealing of that cellular entities personhood.

Even if our prediction for the iPSCs available today holds, and reconstituted blastocysts iPSCs + tTE fail to develop, reprogramming techniques could be improved in the future. As said, fully grown mice have recently been derived from iPSCs, showing that it may be possible to revert somatic cells into a cell entity from which an individual can be born. This means that we now have the capacity to induce totipotency and not just pluripotency. Suppose one could isolate the factors in the tetraploid ICM that are responsible for making the ES, pES, and iPS cells equivalent to the ICM cell layer, that is, capable of generating a full adult body. Then, by using such factors *in vitro* one could reprogram somatic cells not only to iPSCs sharing the potential of ESCs, but to cells sharing the potential of embryonic ICM. This would be the case if the corresponding iPSCs + tTE blastocysts develop to term. Such experiments in animal models have patently a high scientific interest.

I argue now that the proposed experiments using reconstituted ESCs + tTE and iPSCs + tTE blastocysts have a high ethical relevance as well. The achieved generation of adult mice from iPSCs is casting doubts about whether producing iPSCs may become a new sort of cloning (Cyranoski 2009; Denker 2009; Brugger 2010). This presents a dilemma for opponents of human embryo research.

Deducing the moral status of an embryo from its developmental potential is based on the following assumption: that the ICM is *totipotent*, i.e., has the *proper biological potential* to become an adult, and so has a unique moral status; the TE, on the other hand, merely provides the appropriate environment for the ICM to develop into the embryo, the same way as the uterus provides the appropriate environment for the embryo to develop into the adult.

This assumption has been questioned on the basis of the birth of mice fully derived from ESCs and iPSCs. One interprets this result in the sense that the potential of iPSCs or ESCs match that of ICM cells: "ICM cells, iPSCs, and ESCs all require a trophoblast or another suitable substitute to develop" (Devolder 2009), or further one could claim that it "clearly demonstrates that even somatic cells alone have the potential to become an adult human being" (Sagan and Singer 2009). These claims implicitly assume that ICM cells, iPSCs, and ESCs are all equally capable of developing to a full mouse by merely placing them into contact with a trophectoderm (TE) *alone* (Devolder uses the term "trophoblast" instead of "trophectoderm"). But this cannot truthfully be stated without doing the experiment I am proposing.

Suppose, for example, that one produces iPSCs with a certain reprogramming method A. Suppose one can make a mouse from such iPSCs only when these are put in contact with the tetraploid ICM cells, whereas without contact with ICM cells the iPSCs exhibit a potential similar to that of pESCs, i.e., they are capable of reaching the heart beat stage, but not spontaneous fetal motility, like the standard parthenotes. This would be a sign that the iPSCs produced with method A are not totipotent on their own. An organism exhibiting heart beating but incapable of spontaneous motility can be considered to share the status of a brain dead human being. That is, the iPSCs have a DIANA insufficiency and are therefore *not* a human being. Consequently, the method A is ethically unobjectionable for those who oppose embryo destructive research.

Suppose now one improves the aforementioned method to derive a better reprogramming method B. Suppose it is possible to make a mouse from the iPSCs produced with method B *also* when these iPSCs are injected into a tTE (i.e., a blastocyst without ICM cells): this would be a sign that the iPSCs produced with method B are totipotent on their own, and like the ICM layer should be considered a human being. Consequently, method B is ethically objectionable for those who oppose embryo destructive research.

It is clear that if one accepts method B one cannot consistently reject embryo research. Indeed, in accepting method B one implicitly assumes that the corresponding iPSCs acquire totipotency and therefore have a unique moral status only after their interaction with the TE cells (which are tetraploid). But then one should similarly accept that the embryo becomes totipotent, and gets moral status, only after implantation and interaction with the mother's uterine lining. On the other hand, one can consistently accept method A and also reject any destruction of human embryos for research.

To conclude this section, I should want to say that work in progress in the reprogramming of adult somatic cells (Cohen and Melton 2011) shows that, even within the context of cellular reprogramming, it is crucial to have rules that allow us to distinguish between cell entities that are *totipotent*, and therefore constitute human beings, and those that are merely *pluripotent*, and therefore are not human beings. Experiments combining ESCs and iPSCs with tetraploid TE allow us to decide whether the ICM plays a role in making ESCs and iPSCs totipotent and can be done using well known techniques. Such experiments are ethically relevant for deciding which reprogramming methods result in the creation of a new human being and therefore would be inconsistent with the moral requirement of not destroying human embryos.

11.8 Are Humans the Only "Free Agents" in the Universe?

If one assumes that spontaneous movements somewhat involve will and thus directly reveal spiritual agency, it seems to follow that since both human and nonhuman animals exhibit such voluntary movements (Aquinas STh I-II 6.2, Aristotle NE), then both human and nonhuman animals have spiritual powers. That is, if one assumes that human spontaneous movements reveal spiritual agency, it seems that one should consequently assume that the spontaneous movements of nonhuman animals reveal spiritual agency as well.

As we have seen in Sect. 11.2, one would have to accept such a conclusion not only regarding the behavior of animals, but also that of elementary particles all over the universe. Indeed, if one accepts the free will of the experimenter, then quantum physics implies the presence of nonmaterial agency in nature even independently of the working of the human brain. This is a position supported by today's biology as well (Heisenberg 2009).

Quantum physics actually suggests that all phenomena in nature involve nonmaterial agency although at different levels: The nonmaterial control happening in a quantum interferometer is very low, that in a brain very high. From this point of view, "creating" a synthetic bacterium (Gibson et al. 2010) does not essentially prove greater "powers of creation" than arranging an interferometer in a quantum optics lab: In both cases one is setting up appropriate material conditions that allow nonmaterial agency to appear, but in both cases this sort of agency comes from outside space–time and, therefore, is beyond the scientist's creative capabilities. Suppose, as stated in the previous Sect. 11.7, that reprogramming techniques become improved to the extent of creating a human being from a somatic cell. Although a spiritual human soul would certainly appear at the end of the reprogramming procedure, this soul would not be created by the scientist.

But where does the nonmaterial agency in the universe outside "free scientists" come from?

When confronted with such a question, at least three positions seem plausible: *first*, the elementary particles all over the universe and nonhuman animals are animated by a spiritual soul like the human one (e.g., "the free will theorem" by Conway and Kochen 2006, 2009); *second*, the behavior of elementary particles and nonhuman animals requires decisions coming from some ultimate controlling and spiritual being (God); *third*, the behavior of the elementary particles and nonhuman animals require a spiritual agency other than the divine or that which comes from a human-like spiritual soul. In other words, particles and nonhuman animal bodies are not the bodies of some free agent, but are simply acted upon by some immaterial (spiritual) agent other than God (e.g., angels). While this is a very interesting question, it is beyond the scope of this chapter to discuss each of them in detail and even further beyond the scope of this chapter to muse about which position makes the most sense.

11.9 Human and Animal Rights, and Human–Animal Hybrids

The *first* position in the preceding Sect. 11.8 may be related to the widespread tendency of accepting that animals have rights just as humans do. As stated in Sect. 11.2, I conclude that the human body sitting in front of me is a person because

(a) it has the same specific form (or shape) as my body, and (b) this body exhibits movements like the movements I make for expressing my thoughts, emotions, and claims for rights.

In this sense, *human bodily architecture* is a necessary observable condition for defining rights. The best way I can ensure being respected by others is to assume that spontaneous movements in a body of the human species reveal personal agency, and making this assumption the basis of my assigning rights to others. A human body that performs movements like the movements I make to express my feelings and rights-claims is animated by a spiritual soul and therefore is a person I have to respect. Otherwise, I cannot rationally claim that he should presume to respect me. Civil and penal law, for instance, actually assume that the behavior of *human body* is the observable basis for deciding about its rights and culpability. Rights originate from the will to grant that bodies of the human species respect each other.

It is primarily because one wishes to coherently explain human bodiliness and organize human society on the basis of rights that one derives the concepts of soul and life, and one applies them subsequently and somewhat by analogy to *animals*, which are often characterized as organisms capable of spontaneous and voluntary motion. By attributing rights to animals, however, one utterly disposes of the human bodily architecture as an observable basis for defining rights. It is worth asking then whether it is still possible to grant a rational foundation for rights.

Similarly, insofar as the production of human-animal embryonic hybrids blurs the difference between the human species and nonhuman ones it may abolish the possibility of assigning rights. Therefore the procedure would be objectionable even if it did not imply destroying embryos. In this context one can establish the following general rule: the generation of bodies carrying a human brain within a nonhuman animal architecture *and* of bodies with human appearance or architecture carrying a nonhuman animal brain should be prohibited insofar as they distort and undermine an identifiable and stable human bodily architecture, which, as I have argued, serves as the most reasonable foundation for prescribing rights to a particular body. (It seems to me that the Instruction *Dignitas personae* (Congregation for the Doctrine of Faith 2008) is implicitly assuming this rule when it considers unethical procedures "capable of disrupting the specific identity of man").

11.10 Conclusion: *Spontaneous Movements' Metaphysics* Is the Basis for a Consistent Metaphysics of the Human Body–Soul Relation

Distinguishing between embryos and nonembryos, and in general between totipotent and pluripotent (nontotipotent) cell entities, is a goal that hinges on the correct identification of the appropriate biological conditions revealing spiritual ensoulment. I have argued that this can be done on the basis of the distinction between spontaneous and autonomous motions, and the principle that spontaneous movements of a human body directly reveal animation by a spiritual soul. I then showed how it is possible to move from this principle to the *capability* for spontaneous movements as the sign for rational ensoulment of a human body, and followed that up with some proposed criteria as well as the proposal for an experiment to decide when reprogramming of adult somatic cells may produce a human being.

Still further, I endorsed the philosophical position that the *human body's spontaneous behavior* is the basis for deriving the categories allowing us to understand the world and organize society. One cannot appropriately decide when the life of the human person begins and ends on the basis of principles like "movement," "life," "organization," "consciousness," or "thinking." As a basis for arguing and deciding this crucial question, one has to take the body exhibiting human architecture and spontaneous movements as the normative instance of what it means to be a body infused with a spiritual soul.

This *spontaneous movements' metaphysics* is the metaphysics we intuitively often use in our daily life. Each of us is not a pure consciousness, but only a limited (embodied) one. Often I behave without purpose in an unconscious way. But even when I sleep I assume that I will be respected on the basis of my behavior. That is, I act as if everyone can realize that I am a person without the necessity of them awaking me. This wish implicitly involves an adherence to a *spontaneous movements' metaphysics*. By contrast, if I assume that only cognitive functions directly reveal spiritual powers, I cannot assume that while I am sleeping my behavior directly reveals that I am spiritually animated and thus a subject of rights. It is on the basis of spontaneous motility that I immediately conclude that a human body walking on the street or sleeping in a bed is animated by a spiritual soul and is a person with rights.

The progress of medicine has made it possible to transplant hearts and to produce "brain dead" organisms capable of autonomous movements but lacking the capability to make spontaneous ones. These new situations have brought to light something that the intuitive *spontaneous movements' metaphysics* implicitly assumes: Functions characteristic of vegetal life by themselves do not reveal spiritual animation. We can see that the generally accepted clinical criteria for defining death are a general consequence of such an assumption.

I would like to conclude by once again stressing what was shown in Sect. 11.3: that this *spontaneous movements' metaphysics* can be considered a further development of Thomas' reflections on "imperfect voluntary movements." The hylemorphic substantial unity of the human composite understood apart from a *spontaneous movements' metaphysics* causes Thomas Aquinas to accept the (incorrect) hypothesis of delayed hominization. It seems that a consistent metaphysics of the human body–soul relation requires the assumption that *the human body's spontaneous movements directly reveal animation by a spiritual soul*.

Acknowledgments The arguments discussed in this chapter are mainly based on comments by Maureen Condic, Neville Cobbe, Joachim Huarte, Craig Iffland, Juan José Sanguineti and Adrian Walker: I am indebted to all of them. I am thankful to Richard Gardner, Joachim Huarte, Janet Rossant, John West, and Guangming Wu for discussions regarding the experiment proposed in Sect. 11.7.

References

Aquinas Thomas (STh I-II 6.2) Summa theologica I-II. Q. 6.2. http://www.newadvent.org/summa/2006.htm#article2 Cited 10 March 2011

Aquinas Thomas (STh I 76.3) Summa theologica I. Q. 76.3, Reply to objection 3. http://www.newadvent.org/summa/1076.htm#article3 Cited 10 March 2011

Aquinas Thomas (STh I 118.2) Summa theologica I. Q. 118.2, Reply to objection 2. http://www.newadvent.org/summa/1118.htm#article2 Cited 10 March 2011

Aristotle (GA) On the generation of animals. Book II, 3. http://ebooks.adelaide.edu.au/a/aristotle/generation Cited 9 March 2011

Aristotle (NE) Nicomachean Ethics, Book III, 1. The Internet Classics Archive. http://classics.mit.edu/Aristotle/nicomachaen.3.iii.html Cited 26 Feb 2011

Bell JS (1987) Speakable and unspeakable in quantum mechanics. University Press, Cambridge

Boland MJ, Hazen JL, Nazor KL, Rodriguez AR, Gifford W, Martin G, Kupriyanov S, Baldwin KK (2009) Adult mice generated from induced pluripotent stem cells. Nature 461:91–94

Branciard C, Brunner N, Gisin N, Kurtsiefer C, Lamas-Linares A, Ling A, Scarani V (2008) Testing quantum correlations versus single-particle properties within Leggett's model and beyond. Nat Phys 4:681–685. doi:10.1038/nphys1020

Brugger EC (2010) Parthenotes, iPS Cells, and the product of ant-Oar: a moral assessment using the principles of hylomorphism. Natl Cathol Bioeth Q 10:123–142

Cohen DE, Melton D (2011) Turning straw into gold: directing cell fate for regenerative medicine. Nat Rev Gen 12: 243–252

Chen Z, Liu Z, Huang J, Amano T, Li C, Cao S, Wu C, Liu B, Zhou L, Carter MG, Keefe DL, Yang X, Liu L (2009) Birth of parthenote mice directly from parthenogenetic embryonic stem cells. Stem Cells 27:2136–2145

Congregation for the Doctrine of Faith (2008) Intruction *Dignitas Personae*: 33

Conway JH, Kochen S (2006) The free will theorem. Found Phys 36:1441–1473

Conway JH, Kochen S (2009) The strong free will theorem. Not AMS 56:226–232

Counter SA (2007) Preservation of brainstem neurophysiological function in hydranencephaly. J Neurol Sci 263:198–207

Cyranoski D (2009) Mice made from induced stem cells. Nature 460:560

De Vries JIP, Visser GHA, Prechtl HFR (1982) The emergence of fetal behaviour: I qualitative aspects. Early Hum Dev 7:301–322

Denker HW (2009) Ethical concerns over use of new cloning technique in humans. Nature 461:341

Devolder K (2009) To be, or not to be? Are induced pluripotent stem cells potential babies, and does it matter? EMBO Rep 10:1285–1287

Gardner RL, Barton SC, Surani MA (1990) Use of triple tissue blastocyst reconstitution to study the development of diploid parthenogenetic primitive ectoderm in combination with fertilization-derived trophectoderm and primitive endoderm. Genet Res 56:209–222

Gibson DG, Glass JI, Lartigue C, Noskov VN, Chuang R-Y et al (2010) Creation of a bacterial cell controlled by a chemically synthesized genome. Science 329:52–56

Heisenberg M (2009) Is free will an illusion? Nature 459:164–165

Huarte J, Suarez A (2011) Embryos grown in culture deserve the same moral status as embryos after implantation. In: Suarez A and Huarte J (eds) Is this cell a human being. Springer, Berlin, pp 55–75

Kang L, Wang J, Zhang Y, Kou Z, Gao S (2009) iPS can support full-term development of tetraploid blastocyst-complemented embryos. Cell Stem Cell 5:135–138

Kono T, Sotomaru Y, Katsuzawa Y, Dandolo L (2002) Mouse parthenogenetic embryos with monoallelic H19 expression can develop to day 17.5 of gestation. Dev Biol 243:294–300

Kono T, Obata Y, Wu Q, Niwa K, Ono Y, Yamamoto Y, Park ES, Seo JS, Ogawa H (2004) Birth of parthenogenetic mice that can develop to adulthood. Nature 428:860–864

Kuno N, Kadomatsu K, Nakamura M, Miwa-Fukuchi T, Hirabayashi N, Ishizuka T (2004) Mature ovarian cystic teratoma with a highly differentiated homunculus: a case report. Birth Defects Res A Clin Mol Teratol 70:40–46

Laureys S (2005) Science and society: death, unconsciousness and the brain. Nat Rev Neurosci 6:899–909

Libet B (2002) The timing of mental events: Libet's experimental findings and their implications. Conscious Cogn 11:291–299

Meissner A, Jaenisch R (2006) Generation of nuclear transfer-derived pluripotent ES Cells from cloned Cdx2-deficient blastocysts. Nature 439:212–215

Nakamura Y, Takaira M, Sato E, Kawano O, Miyoshi O, Niikawa N (2003) A tetraploid liveborn neonate: cytogenetic and autopsy findings. Arch Pathol Lab Med 127(12):1612–4

Pangallo M (2007) Il pensiero di San Tommaso riguardo all'embrione umano. In: Sgreccia E et Laffitte J (eds) L'embrione umano nella fase del preimpianto: Aspetti scientifici e considerazioni bioetiche. Libreria editrice vaticana, Città del Vaticano. http://www.academiavita.org/index.php?option=com_content&view=article&id=296:m-pangallo-il-pensiero-di-san-tommaso-riguardo-allembrione-umano&catid=57:atti-della-xii-assemblea-della-pav-2006&Itemid=66&lang=it Cited 9 March 2011

Poueymirou WT, Auerbach W, Frendewey D, Hickey JF, Escaravage JM, Esau L, Doré AT, Stevens S, Adams NC, Dominguez MG, Gale NW, Yancopoulos GD, DeChiara TM, Valenzuela DM (2007) F0 generation mice fully derived from gene-targeted embryonic stem cells allowing immediate phenotypic analyses. Nat Biotechnol 25:91–9

Rizzolatti G, Sinigaglia C (2010) The functional role of the parieto-frontal mirror circuit. Nat Rev Neurosci 11: 262–274

Rossant J, Spence A (1998) Chimeras and mosaics in mouse mutant analysis. Trends Genet 14:358–363.

Sagan A, Singer P (2009) Embryos, stem cells and moral status: a response to George and Lee. EMBO Rep 10:1283

Stefanov A, Zbinden H, Gisin N, Suarez A (2002) Quantum correlations with spacelike separated beam splitters in motion: experimental test of multisimultaneity. Phys Rev Lett 88:120404

Stefanov A, Zbinden H, Gisin N, Suarez A (2003) Quantum entanglement with acousto-optic modulators: 2-photon beatings and Bell experiments with moving beamsplitters. Phys Rev A 67:042115

Suarez A (1993) Sono l'embrione umano, il bambino con anencefalia ed il paziente in stato vegetativo persistente persone umane? Una dimostrazione razionale a partire dai movimenti spontanei. Acta Philos 2:105–125

Suarez A, Lang M, Huarte J (2007) DIANA anomalies. Criteria for generating human pluripotent stem cells without embryos. Natl Cathol Bioeth Q 7:315–335

Suarez A (2008) Nonlocal "realistic" Leggett models can be considered refuted by the before-before experiment. Found Phys 38:583–589

Walker A (2005) Resonable doubts. A reply to E. Christian Brugger. Communio 32:771–783

Weiss JR, Burgess JR, Kaplan KJ (2006) Fetiform teratoma (homunculus). Arch Pathol Lab Med 130:1552–1556

Zhao XY, Li W, Lv Z, Liu L, Tong M, Hai T, Hao J, Guo CL, Ma QW, Wang L, Zeng F, Zhou Q (2009) iPS cells produce viable mice through tetraploid complementation. Nature 461:86–90

Glossary

If an author introduces a new concept or different authors give different meanings to the same term, we refer to the chapter where the corresponding definitions appear.

Active potential *noun*, an active potential is one that is actualized wholly from within. It is indicative of an entity's nature – its ontological status. For example, an acorn has an active potential to become an oak tree. See Chap. 4 for usage in context.

Altered nuclear transfer (ANT) *noun*, genetic alteration of the nucleus of a somatic cell (a skin cell for example) before transferring it into an enucleated oocyte. In the most widely discussed example, one inactivates a gene crucial for trophectoderm development. The inactivation eliminates the potential to form the fetal–maternal interface, but spares the inner cell mass lineage. Several authors (Chaps. 2, 3, 4, 6, and 10) assume that the cell entities obtained through ANT would not be organisms and therefore should be considered non-embryos rather than disabled ones. Other authors (Chaps. 5 and 11) conclude that ANT cells may be defective human beings whose development is impaired by a defect that hinders the implantation: To decide the issue one should combine the ANT-derived inner cell mass with a healthy trophectoderm and see whether the so reconstituted blastocyst is capable of reaching the stage showing fetal motility. Finally, there are authors (Chaps. 6 and 7) stressing that as far as there is doubt about the status of ANT cells one should avoid using them for obtaining human pluripotent stem cells.

Androgenote *noun*, a diploid egg cell containing only the paternal genome. Such cells arise sometimes in nature, if for instance there is a loss of the maternal chromosomes, together with a replication of the paternal chromosomes before cell division. Androgenotes carry faulty epigenetic information due to a lack of maternal genes and can produce complete hydatidiform moles in the uterus.

Aneuploid *adjective*, having a chromosome number that is not an integral multiple of the haploid number of the species; an abnormality in chromosome number.

A. Suarez and J. Huarte (eds.), *Is this Cell a Human Being?*,
DOI 10.1007/978-3-642-20772-3, © Springer-Verlag Berlin Heidelberg 2011

Blastocyst *noun*, a hollow, spherical structure during early development of mammalian embryos, comprising an outer layer of cells, called trophectoderm (ultimately giving rise to the placenta and other supporting tissues needed for fetal development) surrounding a small cavity and containing a cluster of cells at one side, called the inner cell mass (which gives rise to the tissues of the embryonic body after implantation). The human blastocyst, which develops about 5 days after fertilization, consists of approximately 70–100 cells.

Blastomere *noun*, any of the cells resulting from the first few divisions of the ovum after fertilization.

Brain-dead *adjective*, the state of a human organism fulfilling the so-called criteria of brain death. A brain-dead body may in fact maintain integrative capacities to a certain degree. If by death one means the "loss of the integrative capacity of the organism", then certain authors claim that a "brain-dead" body is a living organism and should not be declared "dead". Other authors (Chap. 5) suggest that death means the "loss of the integrative capacity for spontaneous motility (animal behavior)", and this is what the standard clinical criteria of brain death actually attempt to ascertain.

Chimera *noun*, a term used either broadly to describe any single biological entity composed of materials from different organisms or more specifically to describe a single organism containing a mixture of genetically distinct populations of whole cells ultimately originating from different embryos of the same or different species; most commonly used to refer to combinations of cells from different species. The name originally designated a mythological fire-breathing monster that was part lion, part goat, and part snake.

Chimeric *adjective*, relating to a chimera or chimeras.

Chromosome *noun*, any of the typically linear bodies in the cell nucleus of eukaryotic organisms, which contain most or all of the genes of the organism and take up basophilic stains (hence the name, literally meaning "colored body"). The term is also applied more specifically to a condensed form of chromatin (DNA and associated proteins) prior to cell division; when chromosomes have replicated but are still attached at the centromere, they are called sister chromatids.

Cognate gene *noun*, a gene that is similar in nature, character, or function to other genes; generally genes that are similar but not identical between species (i.e., homologous).

Complete hydatidiform mole *noun*, a mass of placental tissue having the appearance of a bunch of grapes. Complete moles are supposed to come from *androgenote eggs* when either a single sperm fertilizes an enucleated egg and then undergoes a duplication of its haploid genome or two sperm fertilize an enucleated egg (approximately 20% of complete moles). A type of tumor called "choriocarcinoma" can arise from these entities.

CpG islands *noun*, genomic regions containing a comparatively high frequency of cytosine and guanine and which tend to be associated with the promoters of mammalian genes. Such regions may be hundreds of base pairs long and are typically methylated during phenomena such as genomic imprinting or X-chromosomal inactivation.

Cybrid *noun*, a portmanteau of "cytoplasmic hybrid". The term was originally applied to eukaryotic cell lines produced by the fusion of two distinct cell types, one of which was depleted of its own mitochondria beforehand (by prolonged incubation with ethidium bromide) while retaining its nuclear genome, thereby allowing the nuclear genome from one source to be combined with mitochondrial genomes from other sources. The term has latterly been used to also describe an embryonic entity produced by fusing an enucleated egg with a somatic cell containing the nucleus from a different species.

Cytochrome *noun*, a class of iron-containing proteins typically found in the mitochondrial inner membrane and important for carrying out electron transport during cellular respiration.

Cytoplasmic *adjective*, relating to the cytoplasm of a eukaryotic cell, comprising the semifluid matter contained within the plasma membrane but excluding the cell nucleus.

DIANA insufficiencies *noun (plural)*, anomalies of the genetic and/or epigenetic information that directly inhibit the appearance of neural activity responsible for spontaneous fetal motility.

Diploid *adjective*, the normal state for somatic (i.e., body) cells in mammals, in which the basic (haploid) chromosome number is doubled so each distinct chromosome is paired with its counterpart.

DNA *noun*, abbreviation for deoxyribonucleic acid; a nucleic acid that forms the molecular basis of heredity in cellular entities and localized chiefly in the cell nuclei of eukaryotes. The molecule typically consists of a double helix held together by hydrogen bonds between purine and pyrimidine bases, which project inward from two chains containing alternate links of deoxyribose and phosphate.

Embryo *noun*, an animal during the early stages of growth and differentiation that are characterized by cleavage, the laying down of fundamental tissues, and the formation of primitive organs and organ systems; the term is especially applied to the developing human individual from the time of fertilization (since cleavage commences immediately after fertilization to produce the two-cell embryo) to the end of the eighth week after conception (up to the point of skeletal formation).

Enucleated *adjective*, denoting a cell from which the nucleus has been removed.

Epigenetic *adjective*, relating to a modification in gene expression that does not directly depend on the primary DNA sequence of a gene; generally, modifications in gene expression that are controlled by heritable but potentially reversible changes in DNA methylation and/or chromatin structure.

Epigenetic information *noun*, the pattern of activated and inactivated genes that regulates gene expression during development.

Epigenetic state *noun*, the actual configuration of activated and inactivated genes at a certain developmental stage. As described in Chaps. 3 and 5, the fertilized egg and a parthenogenetic one exhibit different epigenetic states.

Fallopian tube *noun*, either of the pair of tubes that carry the eggs from the ovary to the uterus; also described as the uterine tube.

Fertilization *noun*, the process whereby two haploid gametes unite so the somatic chromosome number is restored and the development of a new individual is initiated.

Gamete *noun*, a specialized and mature male or female reproductive cell (e.g., male sperm cell or spermatozoon and female egg cell or ovum) containing the haploid set of chromosomes, which is typically able to unite with another reproductive cell of the opposite sex during fertilization to form a zygote.

Gametogenesis *noun*, the process of development whereby male or female reproductive cells within the germline undergo meiosis and maturation, leading to the production of gametes.

GC concentration *noun*, the percentage of cytosine and guanine nucleotides in a given polynucleotide sequence.

Gene *noun*, a specific sequence of nucleotides that is usually located on a chromosome that constitutes the functional unit of inheritance by controlling the transmission and expression of one or more traits, either by specifying the structure of a particular polypeptide or controlling the function of other genetic material.

Gene knockout *noun*, a genetically engineered organism that carries a gene (or genes) in its chromosomes that has been made inoperative. Gene knockouts are used to study the biological role of the "knocked out" gene or to produce organisms with specific anatomic, physiological, or behavioral deficits.

Genome *noun*, the total content of genetic material present in a cell or organism.

Genomic imprinting *noun*, an epigenetic process that involves methylation and histone modifications in order to achieve monoallelic gene expression without altering the genetic sequence. These epigenetic marks are established in the germ line and are maintained throughout all somatic cells of an organism.

Germline *noun*, the gamete-producing cells of an organism in which the genetic material may be heritably transmitted through successive generations.

Haploid *adjective*, having the same number of chromosomes found in gametes or half the number characteristic of somatic cells of a given species.

Heterochrony *noun*, an evolutionary phenomenon that involves changes in the rate and timing of species development.

High cervical quadriplegia *noun*, a high-level spinal injury (neck injury) that leaves the patient fully paralyzed and unable to breathe without technical assistance. Christopher Reeve was a famous example. Left alone, such patients ordinarily die. They cannot be "cured", only maintained on life support. In many cases, they even need assistance to maintain a regular heart beat. If support is provided and the person does not die immediately, such patients remain fully conscious and capable of communication through strongly reduced motility mediated by the cranial nerves. However, their brains no longer control the function of their bodies. It is argued that according to the "organization" criterion, the somatic integrative unity of such patients cannot be considered larger than that of a body declared to be brain dead (see Chap. 5).

Histone *noun*, any of a family of highly conserved water-soluble proteins that are rich in the basic amino acids lysine and arginine and are complexed with DNA.

Human being *noun*, the individual of the human species under the aspect that he or she is subject of rights, especially right to life, and deserves the corresponding respect on the part of other human beings. In this sense, human beings can be considered synonymous with "human person".

Human person *noun*, variously defined but often understood as the "individual substance of a rational nature" (definition according to Boethius in Chap. 7). From the perspective of the observable operations or signs allowing us to determine whether a cell entity is a human person, there are two main positions: (a) A cell entity is a human person if it is a human organism (Chaps 3, 4, and 10) (b) A cell entity is a human person if it is a human body exhibiting spontaneous movements or it has the *proper biological potential* for developing the neural activity responsible for such movements (Chaps. 5, 6, and 11). Position (b) links the definition of person to the capability for developing corporal operations similar to those one individual of the human species uses to communicate with other individuals of the human species. This position defines "person" through "relation," and assumes it is a basic category for understanding and explaining the world; that is, one defines animals as living beings exhibiting movements like the human spontaneous ones, rather than defining spontaneous movements as those animals exhibit.

Human soul *noun*, the form that is proper to the human body (as defined according to Aristotle and Thomas Aquinas), viewed as the immaterial integrating principle of a living human organism (see Chap. 6). According to certain authors (Chaps. 5 and 11), the presence of the human soul is *directly* revealed by the integrated and coordinated operations proper to an organism with spontaneous motility (animal behavior), and it can likewise be deduced from the observable biological features ensuring the capability to develop spontaneous movements. Even if the human soul as such cannot be directly accessed by experimental procedures, the body of the human species with spontaneous motility is a visible sign of the soul; it is nothing other than the embodied presence of the soul in space and time. In contrast a brain-dead human, even if it exhibits a certain degree of integrated functioning, still lacks the proper biological potential for performing spontaneous movements and therefore it does not share the moral status of a person.

Hydatidiform mole *noun*, a mass in the uterus composed of enlarged and degenerated chorionic villi, growing in edematous clusters resembling grapes, which ordinarily develops following fertilization of an enucleate egg and either may or may not contain fetal tissue. (See also "complete hydatidiform mole".)

Hybrid *noun*, the result of a genetic cross between parents of different kinds (whether different species, genera, or breeds), or a product of somatic cell fusion that contains components from one or more dissimilar genomes.

Hybrid cloning *noun*, production of a hybrid organism through nuclear transfer by placement of a human somatic cell within the cytoplasm of an enucleated animal egg. See also "Cybrid".

Hybridise *verb*, to crossbreed genetically dissimilar individuals or otherwise produce a hybrid.

Imprinted gene *noun*, a gene whose activity is expressed in a parent-of-origin specific manner. In other words, the expression of imprinted genes depends upon whether they were inherited from an organism's mother or father.

Inner cell mass (ICM) *noun*, the mass of cells inside the blastocyst of a mammalian embryo that is destined to become the structures of the postnatal body. According to certain authors (e.g., Chaps. 5 and 11), the human ICM shares the status of a human being, while for other authors (see Chaps. 3 and 4) it shares the status of human pluripotent stem cells.

In vitro fertilization *noun*, the fertilization of an egg in a laboratory glassware, usually in a shallow dish whereby sperm is mixed with eggs that have been obtained from an ovary, to be followed by introduction of one or more of the resulting fertilized eggs into a female's uterus. Usually abbreviated as "IVF".

Implantation *noun*, the process whereby an embryo attaches to the maternal uterine wall; also referred to as nidation.

Interspecific *adjective*, existing or occurring between different species.

Locked-in syndrome *noun*, the condition in which a patient is conscious but cannot communicate with others because of muscular lesions that make him or her incapable to perform spontaneous movements (total locked-in). Patients with partial locked-in can communicate using the movements of the eyes. Distinguishing total locked-in syndrome and persistent vegetative state is a challenging task object of intensive research in progress.

Membrane *noun*, any of the semipermeable limiting layers of a cell that consist of a fluid phospholipid bilayer with intercalated proteins, e.g., the cell membrane or plasmalemma.

Meiosis *noun*, a type of cell division associated with the production of gametes that results in the number of chromosomes in gamete-producing cells being reduced to half the chromosome number of the parent cell. The process effectively involves a reduction division in which one of each pair of homologous chromosomes passes to each daughter cell, followed by a modified form of mitotic division.

Mitosis *noun*, the commonest type of cell division in eukarytotes, resulting in two daughter cells each having the same number and kind of chromosomes as the parent nucleus. The process typically involves a series of steps consisting of prophase, metaphase, anaphase, and telophase.

Morphogen *noun*, any of various chemicals in embryonic tissue that influence the movement and organization of cells during morphogenesis by forming a concentration gradient.

Morula *noun*, a globular solid mass of blastomeres formed by cleavage of a zygote; typically preceding the blastocyst stage of embryonic development.

Microcephaly *noun*, a neurodevelopmental disorder in which affected individuals have a head circumference less than three standard deviations below the average for their age.

Mirror neurons *noun (plural)*, the nerve cells of the mirror system that unifies action production and action observation, thus allowing the understanding of other's actions from the inside (first-person knowledge). Mirror neurons provide a neurophysiological correlate to the spontaneous movements' criterion.

Mitochondrial *adjective*, referring to mitochondria. The mitochondria (*singular mitochondrion*) are numerous structures found within the cytoplasm of eukaryotic cells, with a smooth outer membrane enclosing a deeply infolded inner membrane and primarily functioning as a "powerhouse" by generating adenosine triphosphate (ATP) through the biochemical process of cellular respiration.

Multipotent *adjective*, describing the potential to form multiple cell types within one germ layer lineage, e.g., adult somatic stem cells such as hematopoietic stem cells.

Myosin *noun*, the commonest protein in muscle cells, responsible for the elastic and contractile properties of muscle; related members of a large family of motor proteins found in eukaryotic tissues.

Neural activity *noun*, the activity produced by the neurons and neuronal networks responsible for determined motor or cognitive achievements. The centers responsible for spontaneous motility (e.g., breathing, eyes, arms, and legs movements) are located in the brain stem.

Neuron *noun*, an excitable cell specialized for the transmission of electrical signals (nerve impulses) over long distances within the body; a nerve cell.

ng-Parthenote *noun*, an entity produced by transferring the nucleus of a nongrowing (ng) primary oocyte (egg cell obtained from a new born mouse) into a fully grown oocyte. The transfer leads to a diploid cell, which contains two sets of chromosomes, both of maternal origin. As described in Chaps. 5 and 11, the artifacts resulting from this fertilization-like procedure develop to reach spontaneous fetal motility and even birth.

Nucleus *noun*, a central organelle in most eukaryotic cells, typically a single rounded structure bounded by a double membrane, containing a nucleoprotein-rich network from which chromosomes and nucleoli arise.

Organism *noun*, an individual constituted to carry on the activities of life by means of organs separate in function but mutually dependent. According to certain authors (Chaps. 5 and 11), there are entities like a brain-dead human body that can be considered an organism to a certain degree but lack the potential for neural activity responsible for spontaneous motility and therefore are not human beings and do not share the moral status of a person.

Oocyte *noun*, a female germ cell produced in the ovary that may undergo meiotic division to form an ovum (the egg), which can be fertilized.

Ovum *noun*, a mature female reproductive cell that has undergone a reduction division and takes the form of a relatively large inactive gamete, providing a comparatively great amount of reserve material and contributing most of the cytoplasm of the zygote after fertilization.

Parthenote *noun*, an egg that has been activated to begin to divide and to develop in the absence of sperm and therefore contains only the maternal chromosomes. Such cell entities (also called standard parthenotes) carry faulty epigenetic information due to a lack of paternally imprinted genes and are not capable of developing fetal motility (Chaps. 5 and 11). Certain authors (Chap. 4) consider that human standard parthenotes are not organisms and therefore do not share the moral status of a person; another (Chap. 3) states that they may be severely defective human organisms and human beings, should it be demonstrated that they have some degree of coordinated development; finally other authors (Chaps. 5, 6, and 11) share the view that a human standard parthenote has the status of a brain-dead organism: that is, it can be considered an organism but is not a human being. See also "ng-Parthenote".

Partial hydatidiform mole *noun*, a cellular entity that is generated when two sperm fertilize a normal egg resulting in a *conceptus* with 69 chromosomes. Partial moles often contain a malformed fetus.

Passive potential *noun*, a passive potential, as described in Chap. 3, is a developmental potential that is actualized from without. It requires the active causal intervention of an external agent in order to be realized. Thus, an acorn only has a passive potential to become a crucifix because it would need the agency of a master craftsman in order to realize this end.

Persistent vegetative state *noun*, the state of patients exhibiting spontaneous movements but (in contrast to locked-in patients) lacking consciousness and the capability to communicate with others.

Personal identity *noun*, description of the way the human person exists in time: the person does not change, even though his or her body and personality can develop in time. For example, a human adult and the entities (fertilized egg, ICM, fetus) from which an adult develops are the same human being and share the same personal identity. For further discussion, see Chap. 11.

Phosphorylation *noun*, a biochemical process that involves the addition of a phosphate (PO_4) group to a protein or a small organic molecule.

Phylogenetic *adjective*, relating to the evolutionary history and diversification of a taxonomic group or species based on inherited features in such organisms.

Placenta *noun*, the vascular organ in mammals that unites the fetus to the maternal uterus and mediates its metabolic exchanges through a rather intimate association of uterine mucosal tissues with those of the chorion and usually allantois. As fingerlike vascular chorionic villi are typically interlocked with corresponding modified areas of the uterine mucosa, the placenta permits exchange of material between the maternal and fetal vascular systems by diffusion but without direct contact between maternal and fetal blood.

Pleiotropic *adjective*, describing circumstances where a single gene controls multiple phenotypic traits.

Pluripotent *adjective*, pertaining to the ability of a stem cell to differentiate into many distinct cell types. In animals, pluripotent stem cells can develop into each of the body's cell lineages (including germ cells), but they cannot develop on their own into an entire organism. Pluripotent cells can be derived from the inner cell mass (ICM) of embryos arising from fertilization or parthenogenesis, and through reprogramming of somatic cells as well. Certain authors assume that the ICM itself is just a colony of pluripotent cells (Chaps. 2, 3, and 4). In contrast, other authors (Chaps. 5 and 11) consider that the ICM has the proper biological potential for developing into an entire living animal and, therefore, the human ICM shares the moral status of a whole human being, while the trophectoderm shares the status of a heart or skin. New experiments combining pluripotent stem (ES or iPS) cells with trophectoderm vesicles (blastocysts after removal of the ICM) are crucial to decide whether the ICM is equivalent to a colony of pluripotent stem cells or not.

Polygenic *adjective*, describing circumstances where a phenotypic trait is controlled by the interaction of more than one gene.

Pronucleus *noun*, the haploid nucleus of a male or female gamete until fusion with the genetic contents of another gamete following fertilization.

Proper biological potential *noun*, as described in Chaps. 5 and 11, this refers to the potential of a cell or cluster of cells (X) to develop into a human body (B) exhibiting spontaneous motility. It is defined according to observable biological features (e.g., genetic and epigenetic information) that permit one to determine if the following two conditions are fulfilled: (1) The neural centers of B responsible for its spontaneous activity derive exclusively from cells contained in X. (2) Equivalent neural centers cannot be derived from the cell clusters or cell layers interacting with X. The cell or cell clusters X and B share the same personal identity. As regards the cell clusters or cell layers interacting with X, they share a similar organic status as that of the heart or skin of B, that is, they can be replaced without loss of personal identity. The inner cell mass (ICM) and the fertilized egg from which the ICM develops have the proper biological potential to develop spontaneous motility and share the moral status of a human being. In contrast, the parthenogenetic ICM does not share such a potential. The authors assume that pluripotent stem cells (ES and iPS cells) do not have this proper biological potential and propose a new experiment to decide this question.

Protamine *noun*, any of various strongly basic proteins of relatively low molecular weight that are rich in arginine and associated especially with DNA in the sperm of various animals.

Quantum nonmaterial agency *noun*, quantum effects, as described in Chap. 11, reveal nonmaterial agency: quantum interferences and correlations come from outside space–time in the sense that they cannot be explained exclusively by stories in space–time, that is, material agency.

Recombinant DNA *noun*, a form of artificial DNA that is created by combining two or more sequences that would not normally occur together through the process of gene splicing.

Reporter gene *noun*, a gene within a gene sequence that is easily observed when it is expressed in a given tissue or at a certain stage of development; generally a transfected gene that produces a signal, such as green florescence, when it is expressed; it is typically included with the transfection of another gene (or genes) to follow the temporal and spatial pattern of expression of that gene.

Reprogramming *noun*, altering the epigenetic state of a nucleus and thereby changing its developmental potency or fate; e.g., during cloning, a somatic cell nucleus is reprogrammed by factors within an oocyte to enter into a state similar to that of a zygotic nucleus, so it becomes capable of supporting a normal pattern of embryonic development.

RNA *noun*, abbreviation for ribonucleic acid, a nucleic acid present in all living cells that contains ribose and uracil as structural components. Among its principal roles in cellular organisms is to act as a messenger carrying instructions from DNA to control the synthesis of proteins.

SCNT *noun*, somatic cell nuclear transfer, a technique involving transplantation of nuclei from body (i.e. somatic) cells to enucleated eggs and thereby permitting cloning of animals such as Dolly the sheep.

siRNA *noun*, small inhibitory RNA, a short sequence of RNA which can be used to silence gene expression.

Somatic *adjective*, pertaining to the body and describing any cell of a multicellular organism other than those of the germ line.

Sperm *noun*, short for ***spermatozoon***, a sperm cell. The mature motile male gamete of an animal, by which the ovum is fertilized, typically having a compact head and a long posterior flagellum for swimming.

Spontaneous movements *noun*, movements of the human body similar to the movements one human person makes for expressing thoughts, emotions, wishes, and rights-claims to other human persons (as described in Chaps. 5 and 11). Even if they are often unconscious and unintentional, spontaneous movements are potentially will-directed movements; that is, they are movements that can always be directed by the will when chosen.

Stem cell *noun*, an unspecialized cell that can continuously produce unaltered daughter cells which may in turn give rise to differentiated cells. Embryonic stem (ES) cells are derived from the inner cell mass (ICM) of an embryo at the blastocyst stage. Embryonic-like stem cells can be derived from a parthenogenetic ICM (pES cells) or through reprogramming of somatic cells to induced pluripotent stem (iPS) cells.

Synkaryon *noun*, a somatic hybrid in which chromosomes from two distinct parental cells (generally from different species) are enveloped in a single nucleus.

Syngamy *noun*, sexual reproduction by union of gametes. The term is commonly used to refer to the breakdown of pronuclear membranes approximately 24 h after sperm–egg fusion in primates.

Teratoma *noun*, a tumor that contains cells derived from the three germ layers that make up the earliest tissues of an embryo. Sometimes, teratomas contain more complex structures including hair, teeth, and bone.

Tetraploid embryo *noun*, an embryo obtained through fusion of the blastomeres in a two-cell stage zygote. The resulting cells contain four times the basic (haploid) chromosome number.

Tetraploid complementation *noun*, a procedure combining pluripotent stem cells with a blastocyst derived from a tetraploid embryo. Living animals produced by tetraploid complementation may consist of cells coming exclusively from the pluripotent stem cells. Experiments combining ES or iPS cells with pure tetraploid trophectoderm vesicles (blastocysts after removal of the tetraploid ICM) have not yet been published.

Totipotent *adjective*, capable of generating all somatic cells of the fetus. Usually, one considers a cell totipotent only if it is capable of generating all the extraembryonic tissues as well: In this sense, the fertilized ovum and each of the two blastomeres resulting from the first cell division is a *totipotent cell*. However, the morula as a whole can be considered a *totipotent cell entity* in the sense that it is capable of generating a living fetus. As regards the inner cell mass (ICM), certain authors (e.g., Chaps. 2, 3, and 4) consider that it is equivalent to a cluster of pluripotent stem cells (like ES cells). Other authors (Chaps. 5 and 11) consider that the ICM *as a whole* has a higher developmental potential than a cluster of pluripotent stem cells: They claim that ES and iPS cells combined with pure trophectoderm vesicles (without ICM) are not capable of reaching fetal motility. In this sense, these authors consider that the ICM *as a whole* is a *totipotent cell entity* as well, even if the ICM contributes to some extraembryonic tissues but not to all of them (see Chaps. 5 and 11 for further discussion).

Transcription *noun*, the process of constructing a messenger RNA molecule using a DNA molecule as a template, whereby genetic information encoded by a sequence of DNA nucleotides is transferred to the messenger RNA.

Translation *noun*, the process of forming a protein molecule at an organelle within the cell known as the ribosome, whereby a sequence of nucleotide triplets in a messenger RNA molecule gives rise to a specific sequence of amino acids.

Transgenic *adjective*, denoting an organism into which genetic material from an unrelated organism or from another species has been artificially introduced into the germ line, so it can be transmitted from one generation to the next in a manner that ideally ensures its function.

Trophectoderm *noun*, the outer layer of the mammalian blastocyst as distinct from the inner cell mass, supplying the embryo with nourishment and later forming the major part of the placenta.

Trophoblast *noun*, essentially synonymous with trophectoderm. The term is used to denote the tissues into which the trophectoderm differentiates after implantation (e.g., in Chap. 4): the trophoblast develops into placental structures.

Unipotent *adjective*, having the potential to form only one differentiated cell type, e.g., spermatogonial stem cells

Ventricle *noun*, one of a system of four connected fluid-filled cavities within the brain that are continuous with the central canal of the spinal cord.

Xenograft *noun*, a surgical graft of tissue from one species to an unlike species; for example, a graft from a baboon to a human.

Zona pellucida *noun*, the thick transparent membrane surrounding a mammalian ovum before implantation, composed of comparatively elastic noncellular glycoprotein.

Zygote *noun*, a diploid cell resulting from the fusion of two haploid gametes; the term is specifically applied to a fertilized ovum and, more broadly, to the developing individual produced from such a cell.

Index

W

X

Y

Z